T0328920

Paleostress Inversion Techniques

Paleostress Inversion Techniques
Methods and Applications for Tectonics

Christophe Pascal

Professor of Structural Geology, Ruhr University Bochum, Bochum, Germany

ELSEVIER

Elsevier
Radarweg 29, PO Box 211, 1000 AE Amsterdam, Netherlands
The Boulevard, Langford Lane, Kidlington, Oxford OX5 1GB, United Kingdom
50 Hampshire Street, 5th Floor, Cambridge, MA 02139, United States

Notices
Knowledge and best practice in this field are constantly changing. As new research and experience broaden our
understanding, changes in research methods, professional practices, or medical treatment may become
necessary.

Practitioners and researchers must always rely on their own experience and knowledge in evaluating and using
any information, methods, compounds, or experiments described herein. In using such information or methods
they should be mindful of their own safety and the safety of others, including parties for whom they have a
professional responsibility.

To the fullest extent of the law, neither the Publisher nor the authors, contributors, or editors, assume any liability
for any injury and/or damage to persons or property as a matter of products liability, negligence or otherwise, or
from any use or operation of any methods, products, instructions, or ideas contained in the material herein.

Library of Congress Cataloging-in-Publication Data
A catalog record for this book is available from the Library of Congress

British Library Cataloguing-in-Publication Data
A catalogue record for this book is available from the British Library

ISBN: 978-0-12-811910-5

For information on all Elsevier publications
visit our website at https://www.elsevier.com/books-and-journals

Publisher: Candice Janco
Acquisitions Editor: Peter J. Llewellyn
Editorial Project Manager: Emily Thomson
Production Project Manager: Vijayaraj Purushothaman
Cover Designer: Matthew Limbert

Typeset by STRAIVE, India

Working together
to grow libraries in
developing countries

www.elsevier.com • www.bookaid.org

Dedication

To the man who aroused my curiosity, my father, Jacques Pascal.

Contents

Acknowledgements

The writing of a book is never a lonely venture. I am deeply indebted to Damien Delvaux, Olivier Lacombe, Frantz Maerten, François Renard, Aline Saintot and Atsushi Yamaji for their kind help and insightful comments and corrections. I am grateful to the Elsevier team for facilitating the production process and, in particular, to the editorial manager, Emily Joy Grace Thomson, for her infinite patience. A special thanks goes to my family and friends: many thanks for your constant support and encouragement.

1

Introduction

Paleostress inversion refers, traditionally, to mathematical or graphical inversion of geometrical attributes of natural fractures (commonly faults) with the aim of quantifying (most often partially) the stress associated with nucleation or activity of these fractures in the geological past. The definition reveals from scratch that the corresponding research field, namely 'paleostress analysis', is located at the crossroads between structural geology and continuum mechanics. Like many other disciplines that evolved from classical field geology, paleostress analysis attempts to bridge observation of the nature and mathematics. The move follows the general quantification trend of many scientific areas, although the path to walk is often rough and sinuous, especially for geological sciences that are more rooted in observation than in experimentation.

Reconstruction of paleostresses from measurement of natural fractures was originally proposed by Anderson (1905) in his seminal paper. Anderson identified very early the analogy between the geometrical configurations of tectonic faults and Coulomb's theory of rupture. This early attempt to link natural fractures to mechanics found no continuation during the following decades, until Wallace (1951) and Bott (1959) settled the theoretical foundations of fault slip inversion methods in the 1950s. Once again, time passed before Carey and Brunier (1974) set the ground for the very first computer-based paleostress inversion method.

Almost half a century after the historical contribution of Carey and Brunier, an uncountable number of scientific papers dealing with paleostress reconstructions have been published, topical symposiums are frequently organised in geoscience congresses and tens of computer programs for paleostress determinations have been created, many of them being freely accessible on internet. Paleostress analysis has become a very popular tool in structural geology, as suggested by the high numbers of hits that internet search engines return (e.g. ~100,000 for 'paleostress analysis' itself) and, sign of times, as shown by the existence of Wikipedia pages covering the topic.

The research field has gained in maturity with the refinement of the numerical techniques and the empirical demonstration of the validity of the methods, which was conducted by means of applications to natural cases (e.g. Bergerat, 1987; Delvaux and Barth, 2010) and of numerical testing of the background assumptions (e.g. Pollard et al., 1993; Lejri et al., 2017), in parallel. The field has also diversified with the development of innovative paleostress determination methods, based on measurements of tensile fractures (e.g. Jolly and Sanderson, 1997; Yamaji, 2016) or stylolithes (e.g. Schmittbuhl et al., 2004; Ebner et al., 2009), or employing refined geomechanical methods (e.g. Maerten et al., 2016a,b), for example. Clearly, paleostress analysis is nowadays a vivid field of science.

Ironically, I was writing this book when José Simón, a prominent actor in paleostress research, published a note (Simón, 2019) where one could read: "the fact that fault–slip analysis [*i.e. paleostress reconstruction from inversion of fault slip data*] has been very poorly treated in Structural Geology textbooks (with rare exceptions such as Ramsay and Lisle, 2000; Fossen, 2016) speaks against its consideration as a 'mature' discipline."

To my opinion, paleostress analysis is a 'mature' discipline, yet many of its branches are still in an early phase of development, as it will be discussed in this book. In addition, the discipline has considerably diversified since the pioneer works carried out in the 1970s. It was becoming timely, not to say urgent, to summarise the vast amount of knowledge that has been generated during the past decades. The main objective of the present book is to compile the most significant parts of that knowledge and, hopefully, to stimulate further research. The reader will certainly remark on the long list of references I consulted to write the text. I tried to embrace, as much as possible, the information I judged relevant but could not include all paleostress research works in detail. I apologise in advance to readers who, perhaps, will feel that some crucial aspects are missing here.

The book is organised in 10 chapters, including this brief introduction. The journey starts with descriptions of the natural objects under scope (i.e. Chapter 2), that is, tectonic fractures and stylolithes, which represent the fundamental data to the analyses. The following chapter (i.e. Chapter 3) recalls classical notions of continuum mechanics, which are in general involved in the theoretical backgrounds of inversion methods. Chapter 4 discusses extensively the most classical and mature paleostress reconstruction methods, namely fault slip inversion methods. The recent methods that consider inversion of tensile fractures and the more classical methods of inversion of calcite twins are addressed in Chapters 5 and 6, respectively. The following chapters will present alternative and complementary approaches to cope with 'imperfect' data (Chapter 7) and to increase the number of reconstructed stress parameters (Chapter 8). Finally, we will return to the field with some interesting examples of application of paleostress analyses (Chapter 9) and practical advice on how to conduct successfully paleostress studies (Chapter 10).

2

Brittle structures in the field

2.1 Introduction

Paleostress reconstruction methods require in general field measurement of tectonic brittle structures (i.e. tectonic fractures). Therefore the quality of the results depends primarily on the geologist's skills to identify and measure fractures, to interpret their kinematics and to separate them according to their respective chronologies.

Classical paleostress reconstruction methods are mainly based on the measurement of fault planes, including the fault slip indicators they often exhibit (see Chapter 4). More recent methods consider e.g. the spatial distribution of extensional fractures (see Chapter 5). Nevertheless, all kinds of tectonic fractures furnish valuable information at different degrees. Field data are ultimately used in the determination of (1) the types of paleostress regimes that affected the study area (i.e. normal, reverse or strike-slip, see Section 4.1.1), (2) the relative chronologies between these distinct paleostress regimes and potentially (3) their respective timings.

The main objectives of the present chapter are (1) to describe the tectonic objects in the field, which are used for paleostress reconstructions, and (2) to guide the reader in their interpretation. Firstly, fracture classification and formal definitions are introduced. Thereafter, some classical criteria used to determine the sense of slip along faults are given and useful chronology criteria are presented. Finally, field data writing conventions used in this book and elsewhere are described.

2.2 Classification of brittle structures

2.2.1 Brittle deformation and fractures

Brittle deformation of an object refers to deformation promoting loss of continuity of or within the object. The discontinuities eventually created during brittle deformation are called fractures. In contrast to ductile deformation, which implies internal continuity of the object after deformation, brittle deformation occurs mostly in the upper structural levels of the crust.

We will use the word *fracture* to characterise a brittle structure without any other precision, i.e. whenever details on the relative displacements of the fracture walls cannot be assessed. As such 'fracture' is used as a general term for all kinds of brittle structures. Fractures are in turn grouped in *mode I, II and III fractures*, depending on the relative motions of their fracture walls (Fig. 2.1). Mode I (i.e. *opening mode*) applies to extension fractures and relative displacement occurs perpendicular to fracture walls. Mode II

Paleostress Inversion Techniques. https://doi.org/10.1016/B978-0-12-811910-5.00006-3

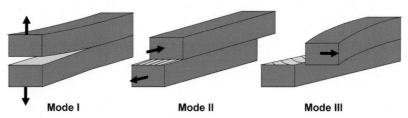

Mode I   Mode II   Mode III

FIG. 2.1 The three modes of brittle deformation or fracture: mode I=opening, mode II=sliding, mode III=tearing.

(i.e. *sliding mode*) and mode III (i.e. *tearing mode*) describe motions along shear fractures and displacement remains parallel to fracture walls by definition. In the case of mode II, a simple translation affects the blocks on either side of the fracture, whereas for mode III the blocks rotate according to an axis perpendicular to fracture walls.

2.2.2 Extension fractures

Extension fractures involve, by definition, separation of fracture walls without relative shear displacement between them, or conversely the displacement vector is perpendicular to fracture walls (i.e. mode I fracture, Fig. 2.1A). Most common tectonic mode I fractures are *extension (or dilatational or tension) joints, extension veins and dykes* (Fig. 2.2). Assuming low strain during the fracture event, the minimum principal stress axis, σ_3, can be taken as nearly perpendicular to the fracture walls. This geometrical relationship implies that the two other stress axes are in the fracture plane but gives no further details concerning their precise orientations a priori. However, we will see in Chapter 5 that fluctuations in orientation of a family of coeval extensional fractures can be used for more accurate stress determinations.

a) Joints

Extension joints (Fig. 2.2) are barren brittle structures found in all kinds of geological environments. The term 'joint' was first coined in the 18th century to suggest that the broken pieces of the rock were joined together across the fractures. The term itself indicates that joint apertures are very modest, i.e. mm scale in general. Their typical lengths range commonly from various metres to 100s of metres, implying length-to-width (i.e. aspect) ratios of 10^3–10^5, which are typical of mode I fractures and in agreement with predictions from linear elastic theory. Extension joints occur often in patterns of systematic and non-systematic joints. Systematic joints are straight, relatively long, and exhibit regular spacing, in particular when they are perpendicular to stratified rocks (Fig. 2.2). Non-systematic joints develop (sub)perpendicular to the systematic ones and in between them. They are shorter and show more erratic trajectories than the members of the systematic set. Systematic joints are believed to reflect regional stress orientations, whereas non-systematic joints respond to stress relaxation following the creation of the systematic set. However, it has been observed (e.g. at Llantwick Major, Wales, UK) that both joint sets can swap their respective orientations while crossing sedimentary interfaces, suggesting that this latter interpretation has to be regarded with some caution.

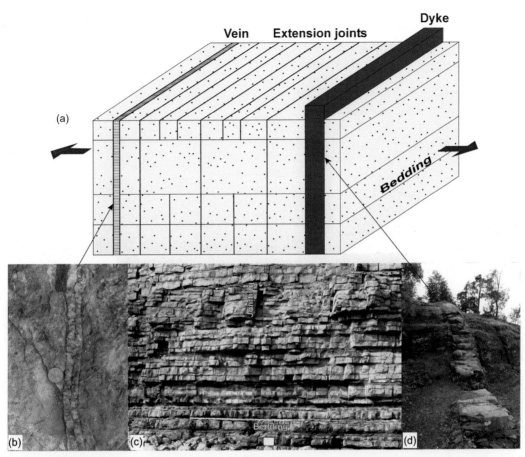

FIG. 2.2 (A) Simplified block diagram depicting the relationship between natural mode I fractures and direction of extension; (B) calcite vein in Upper Triassic meta-limestones (Sierra de Orihuela, Spain); (C) extension joints in Lower Jurassic limestones (Llantwick Major, Wales, UK); and (D) basaltic Permian dyke intruding Cambro-Silurian sediments (Kalvøya, Norway).

Extensional joint faces can be smooth but present very often delicate feathery ornaments (Fig. 2.3), termed plumose (from the French word 'plume' meaning 'feather'). A detailed account on plumose is beyond the scope of this book. We should however note that plumose and related structures (e.g. plume axis, twist hackles) inform on the initiation and propagation of the fracture in the rock mass under (opening) mode I conditions. The presence of plumose on fracture faces indicates unambiguously that the fracture is an extensional joint.

Joints are not obligatory mode I fractures. When two joint sets make an acute angle of ~60°, either the relationship is fortuitous (i.e. the two sets are not coeval) or they were formed in response to shear (see Anderson's theory in Section 4.1.1) and, consequently, the term 'shear joints' should apply. A thorough field analysis and in particular the measurement of compatible structures, like faults, can help to discriminate which case is the

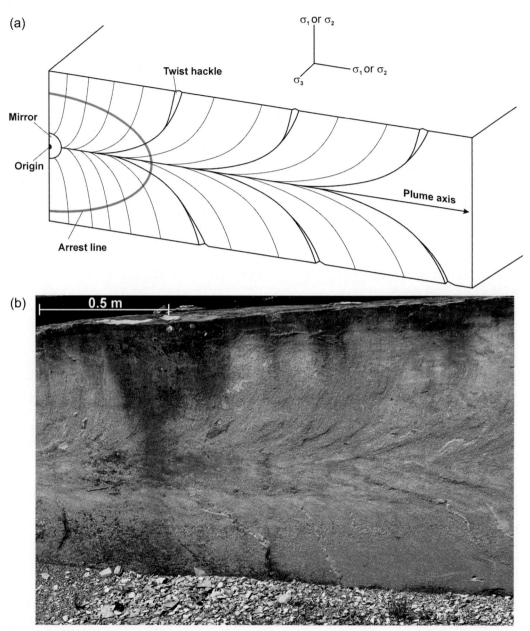

FIG. 2.3 (A) Schematic representation of a joint face with plumose and some of the main structures related to the plumose, the arrow depicts the propagation direction of the joint and (B) example of plumose (extension joint in Upper Cretaceous glauconitic sandstones, Klieve quarry, western Germany).

most likely. No general consensus on the terminology has been reached yet and many authors reserve the word 'joint' to mode I fractures and use 'fault' as a generic name for all kinds of shear fractures (i.e. mode II or III). Hereby, 'joint' is used for fractures without any visible offset parallel to the fracture walls, whether they are nucleated in extension or in shear. This definition might of course be scale-dependent and microscope investigations of thin sections can evidence offsets that remain invisible with the bare eye.

b) Veins

Veins (Fig. 2.2) are extension fractures filled with minerals. The fact that a void had to be created before mineralisation indicates that vein walls moved apart and, subsequently, that veins display characteristics of mode I fractures. In more detail, one has to distinguish between extensional veins, where vein wall displacements occurred strictly perpendicular to the fracture plane, and *hybrid veins*, where the displacement vector was oblique to fracture walls. Assuming low strain, vein-opening directions measured in the field are considered to be (nearly) parallel to σ_3.

Vein width varies commonly from millimetres to decimetres but can reach occasionally more than 1 m. Vein length ranges in general between centimetres and 10s of metres and reaches occasionally 100s of metres. As expected for mode I fractures (or quasi-mode I ones in the case of hybrid veins), aspect ratios are relatively high.

Veins contain usually calcite or quartz, which are the most common minerals, but other mineral species can be found. Precipitation of minerals strongly depends on pressure–temperature conditions, geochemical composition of the circulating fluids and the presence of aliquots. These minerals can provide many important constraints for tectonic analysis. For instance, chemical variations among minerals precipitated successively in a single vein evidence distinct rupture events and, potentially, distinct tectonic events. This first-order inference can find further support if alterations in vein orientation are also associated with changes in mineralogy. The minerals can be dated using analytical methods and, in turn, absolute ages can be devised for the related rupture/tectonic event. We will see in Chapter 8 that studies of fluid inclusions in vein minerals open the way to more complete paleostress tensor reconstructions.

Depending on the interplay between opening and precipitation rates, mineral texture can vary from granular equant (high opening and low precipitation rates, respectively) to fibrous (low opening and high precipitation rates, respectively). Fibrous textures are particularly informative for tectonic analysis, the fibres being parallel to the direction of opening of the vein and, presumably, to σ_3. However, straight fibres represent the simplest case and more complex geometries (e.g. sigmoidal fibres) evidence gradual changes in opening direction (see Ramsay and Huber, 1987, for a detailed account on vein analysis).

c) Dykes and sills

The words '*dyke*' and '*sill*' stem from the respective geomorphological signatures these structures use to leave in the landscape. In other words, dykes are steeply inclined to vertical, whereas sills are horizontal to moderately inclined. A more formal definition states that sills are conformable with the layering of the rock they intrude and that dykes cut through the layering. There is however a marked tendency in the geological community to name these structures according to inclination without any other kind of consideration.

Dykes (Fig. 2.2) and sills are either mode I or hybrid fractures and, thus, the opening directions of their respective walls inform directly on σ_3 orientation. They contain material previously injected between fracture walls at relatively high pressures. The trapped material can be of magmatic or sedimentary (i.e. *clastic dykes and sills*) in origin and may be used to date the tectonic event related to the injection and formation of the structure. Typically, these structures display widths ranging from centimetres to 10s of metres and lengths up to 100s of kilometres (e.g. Cleveland Dyke, UK) and therefore high aspect ratios.

2.2.3 Faults

Faults are mostly shear fractures and, thus, exhibit offsets parallel to fracture walls but comparatively small openings. They are mode II or III fractures (Fig. 2.1B and C), although the latter mode, corresponding to rotational faults, is less frequent in nature than the former. Faults are (quasi)planar objects, framed by an ellipse, as a first-order approximation (i.e. the so-called fault plane) but, in finer detail, fault surfaces are complex geometrical objects. Nevertheless, measurement of average fault orientation in the field (see Section 2.5) determines fault attitude with sufficient accuracy in the case of paleostress reconstructions. Typical fault dimensions range from centimetres to 100s of kilometres and cumulative offsets depend on fault length or, in other words, the longer the fault, the larger the total offset.

A fault represents a geological discontinuity that separates an upper block, the *hanging wall*, from a lower one, the *footwall*. Note that this definition applies only for non-vertical faults. The two terms stem from the British mining sector. When the fault used to crop out on the ceiling of a mine gallery, the so-called hanging wall of the fault was visible above the miner's head. When the fault was intersecting the floor, the 'footwall' was exposed below the feet of the workers. Geologists classify faults according to type of motion or conversely fault slip sense (Fig. 2.4). A *normal fault* (Figs. 2.4 and 2.5A) is a *dip-slip* fault, i.e. fault slip parallels the dip line of the fault plane, whose hanging wall moved downwards relative to the footwall. A *reverse fault* (Figs. 2.4 and 2.5B) is a dip-slip fault whose hanging wall block moved upwards relative to the footwall block. A *strike-slip fault* (Figs. 2.4 and 2.5C) is a fault with horizontal fault slip (i.e. parallel to fault strike). Two senses of slip exist: *dextral* or *right-lateral* strike-slip and *sinistral* or *left-lateral* strike-slip. Note that the terms 'dextral' (i.e. 'to the right') and 'sinistral' (i.e. 'to the left') are viewpoint dependent: one must consider strike-slip faults as they were seen from above. Faults are often *oblique slip*, i.e. they display both dip-slip and strike-slip components. We will see in Chapter 4 that oblique-slip faults contribute significantly to improving the quality of paleostress inversion results. Two special cases of faults are *horizontal faults and vertical faults* (i.e. vertical fault surfaces with significant vertical components of slip). These two types are, at least at the mesoscale, less frequent in the field and reflect local accommodation mechanisms and, hence, local stress reorganisations.

2.2.4 The special case of stylolithes

Stylolithes resemble mm-to-cm-scale sharp blades (i.e. *teeth or stylolitic peaks*), in 3D (Fig. 2.6A) and stock market charts in profile view (Fig. 2.6B). They are found along

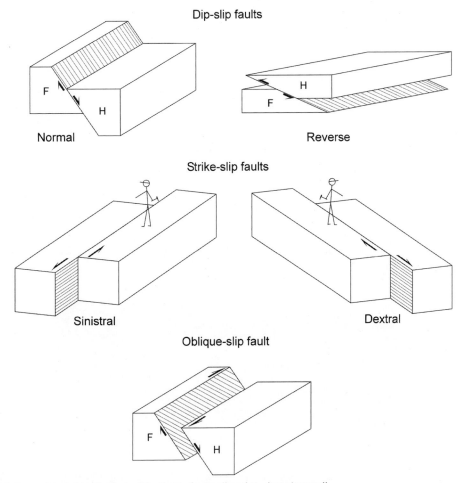

FIG. 2.4 Geometrical classification of faults. F=footwall and H=hanging wall.

(quasi)planar surfaces (i.e. *stylolitic seams or planes*). They are pressure-solution structures resulting from chemical dissolution of rock under relatively high stresses and subsequent transfer of the dissolved minerals to the leaving fluids (see also Section 8.4). The
teeth contain inclusions of insoluble material. In the brittle deformation field, pressure
solution affects mainly rocks involving sufficient amounts of calcite. The process implies
loss of mass but, rigorously speaking, there is no loss of cohesion in the mechanical sense
and the author is of the opinion that stylolithes should not be classified as fractures, as it
has been recently proposed elsewhere (i.e. 'mode IV or closing mode fractures'). Nevertheless, the average orientation of stylolitic peaks parallels the axis of maximum principal
stress, σ_1.

FIG. 2.5 Examples of (A) a normal fault in Upper Carboniferous sediments (Arnao, Asturias, Spain), (B) a reverse fault in Permo-Triassic sediments (Xivares Beach, Asturias, Spain) and (C) (sinistral) strike-slip fault offsetting a calcite vein in Upper Cretaceous limestones (La Murta Valley, Alicante, Spain). *Paired arrows* depict senses of motion along the faults and single arrows indicate markers offset by the faults.

FIG. 2.6 Stylolithes (A) in 3D (Amadorio, Alicante, Spain) and (B) in cross-section (facade wall in Dijon, France).

2.3 Faults and criteria of displacement

2.3.1 Structural criteria

Faults and their associated slips are efficiently detected in the landscape or at outcrops when they offset geological markers (Fig. 2.7A and B). For example, in sedimentary rocks faults disturb bedding continuity and create *stratigraphical offsets* (Fig. 2.7A). However, large fault slips might render difficult the matching of any kind of geological marker, including stratigraphical ones, from either side of a fault. In addition, identifiable markers can be absent in the studied rocks, e.g. in granites. We will see in Chapter 4 that, based on Anderson's theory, fault dip and/or the presence of conjugate faults can furnish a hint on the sense of motion (i.e. on the type of fault), but not all faults are Andersonian and in case of fault reactivation Anderson's theory is of no help (see details in Chapter 4).

Alternatively, one might detect and measure structures resulting from local deformation processes, directly connected to the faulting, and use the gathered information to infer the sense of motion of the fault under scope. A wealth of such structures has been identified and is described in classical textbooks. These structures are important for tectonic analysis but furnish only part of the information needed for paleostress reconstructions. We will therefore restrict their presentation to some of the most classical structures and invite the reader to consult the excellent accounts given in e.g. Twiss and Moores (2007) and Fossen (2016) for further details.

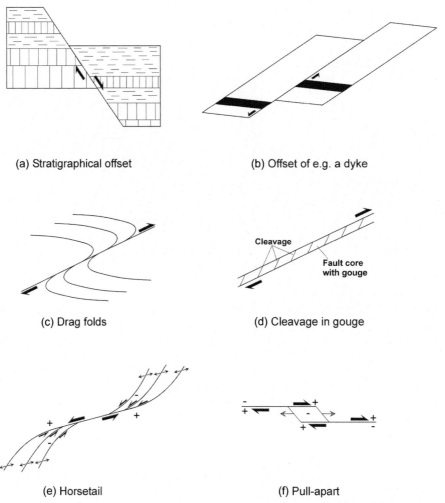

(a) Stratigraphical offset   (b) Offset of e.g. a dyke

(c) Drag folds   (d) Cleavage in gouge

(e) Horsetail   (f) Pull-apart

FIG. 2.7 Some classical criteria of fault displacement; paired and divergent arrows represent fault slip sense and opening directions, respectively. In (E) and (F), pluses and minuses depict, respectively, compressive and tensile sectors near fault tips.

Drag folds (Figs. 2.7C and 2.8A and B) associated with faults are relatively common in all kinds of lithologies and especially in layered rocks. Drag folds form in response to friction on the fault surface. The layers appear like being dragged along the fault surface in a direction opposite to fault slip, because friction resists their motion. As a result, the convex side of the drag fold indicates the direction of motion of the block containing the fold (Fig. 2.8A and B).

Fault cores can potentially involve *gouge* (i.e. fine-grain fault rock resulting from cataclasis of the protolith), which in turn might have developed cleavage perpendicular to the local flattening axis. The orientation of this cleavage informs directly on the sense of

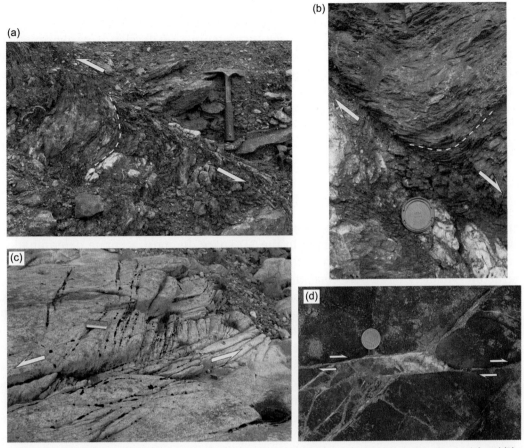

FIG. 2.8 Drag folds associated with (A) a reverse fault in Devonian limestones (Arnao Beach, Asturias, Spain) and (B) a normal fault in Silurian shales (Llumeres Beach, Asturias, Spain), dashed lines indicate the drag folds; (C) horsetail structure at the tip of a sinistral strike-slip fault dissecting Archean gneisses (Lofoten Islands, Norway), the fractures related to the horse tail are particularly well outlined by the vegetation; and (D) pull apart developed in the relay zone of two dextral strike-slip faults in Permo-Triassic sandstones (Xivares Beach, Asturias, Spain), the pull apart was filled with precipitated quartz minerals in the present case. Paired arrows depict senses of motion along the faults in all the pictures.

motion along the fault, the acute angle between cleavage and fault wall opening to a direction opposite to the sense of motion of the corresponding fault block (Fig. 2.7D).

Faults are not infinite, as a consequence any increment of fault slip must be accommodated by deformation and subsequent structures where the fault ends, at its tips and their close vicinities. Depending on their specific locations, these structures are formed either in tension, where the fault block moves away, or compression, where the fault block moves to (Fig. 2.7E). *Horsetails* (Fig. 2.8C) are probably the most common representatives of this family of structures. A horsetail consists of a splay of hybrid fractures (very often referred to as *horsetail splay*), curving away from the main fault towards the receding fault block.

The fractures of the splay show both opening and shear, the latter becoming prominent as the fractures curve to reach a nearly parallel attitude to the main fault.

Fault tips are hence places where a variety of structures can develop, in particular when the tips of two distinct faults are close enough, and the respective deformation patterns of the two faults interact constructively in a region referred to as *relay zone*. It is in such context that a *pull apart* can form (Figs. 2.7F and 2.8D). Pull-apart structures appear as gaps, later filled with e.g. minerals, reflect opening and thus require formation under tension. The kinematics of the pull-apart structure has to be consistent with the kinematics of the faults and, therefore, informs directly on the sense of fault slip. For example, the opening of the pull apart shown in Fig. 2.7F is possible only if the two faults show dextral strike slip.

2.3.2 Slickenside kinematic indicators

In most cases, however, the motion can be assessed only in plane view and the fault slip vector cannot be totally determined in the 3D space. Its most accurate determination requires access to the fault surface itself and the observation and measurement of specific structures, namely *slickenside kinematic indicators*.

In response to shearing, fault surfaces present commonly smooth polished surfaces, termed *slickensides* (Fig. 2.9A) and, in general, slickensides display shear lineations developed parallel to fault slip (*slickenlines or striae*, Fig. 2.9B). They result from 'scratching' of the fault surface by comparatively hard elements in the rock during shear. If the striae present widths larger than few millimetres (Fig. 2.9C), the term 'groove' is preferred. Striae are orientated parallel to the slip vector. However, the information furnished by the measurement of fault striae is restricted to the orientation of the line supporting the fault slip vector. In other words, it is not possible from the measurement of striae alone to discriminate which one of the two possible slip senses along this line is the actual one.

The use of additional kinematic indicators can help to remove the ambiguity. Previous authors have identified numerous slickenside kinematic indicators but many of these are merely rare or difficult to identify in the field. Only the most common and useful ones are described in detail here (Fig. 2.10). The reader is referred to Petit (1987), Angelier (1994) and Doblas (1998) for more complete accounts on the topic.

Slickenside kinematic indicators can be classified in two main categories: (1) structures whose origin involves both mechanical and chemical processes and (2) structures created mainly in response to mechanical processes, i.e. *secondary fractures*. The first category includes *mineral steps* and *stylolitic peaks* or *slickolites* (Fig. 2.10A and B).

Mineral steps are very often observed on slickensides. Most commonly, they are made of calcite or quartz but other minerals may also crystallise, depending on the chemistry of the fluids circulating in the fault zone and on pressure and temperature conditions. The typical dimensions of mineral steps range from millimetres to centimetres. Their mineral fibres develop incrementally during successive fault slip and fill cavities of the fault plane that are protected against abrasion (i.e. *releasing bends*, Figs. 2.10A and 2.11). Accordingly, mineral fibres grow parallel to fault slip and mineral steps point to the direction of motion

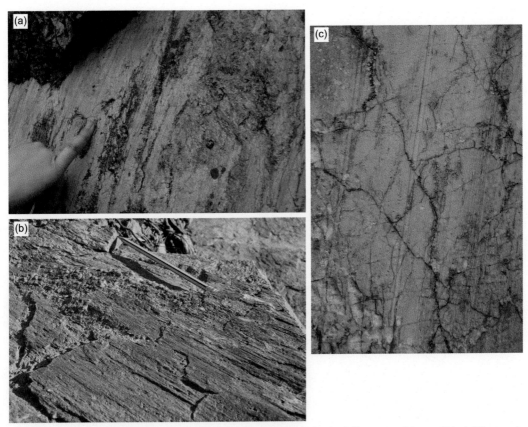

FIG. 2.9 (A) Slickenside presenting a remarkably smooth surface in Palaeozoic limestones (Henan, China); (B) example of striae on a fault plane in Triassic limestones (Sichuan, China); and (C) cm-wide grooves along the Cala de Moraig Fault surface offsetting Upper Cretaceous limestones (Alicante, Spain).

of the missing block (i.e. the fault block removed by erosion). Mineral steps are highly reliable kinematic indicators.

Stylolitic peaks (or *slickolites*) are pressure-solution structures (see Section 2.2.4) formed in zones of the fault surface submitted to pressure during fault slip (i.e. *restraining bends*, Figs. 2.10B and 2.12). As such, stylolitic peaks develop parallel to fault slip but point towards the direction opposite to the displacement of the missing block. These latter kinematic indicators are, by nature, restricted to rocks containing significant carbonate phases. Noteworthy, pressure solution being a relatively long-lasting process, the presence of stylolitic peaks on faults suggests slow creep events.

The second category of slickenside kinematic indicators involves fractures intersecting the surface of the main fault (*M-surface*, Fig. 2.13), formed coeval to it but comparatively less extended. These *secondary fractures* were originally described in 'shear table' experiments by Cloos (1928) and Riedel (1929) and further studied by Tchalenko (1970). They are in the field more ambiguous kinematic indicators than mineral steps or slickolites, and

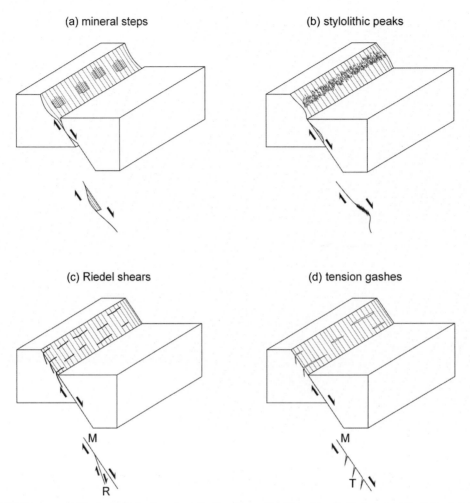

FIG. 2.10 Commonest kinematic indicators along fault surfaces; M=M-surface, R=R-shear and T=T-fracture (or tension gash).

fractures pre- or post-dating the M-surface but intersecting it can easily be misinterpreted as secondary fractures.

Riedel shears (*R-shears*) and *tension gashes* (*T-fractures*) are the commonest secondary fractures observed in the field or, at least, the easiest to identify (Figs. 2.10 and 2.13). R-shears intersect the M-surface at small angles of 5°–25°. They are subsidiary faults presenting senses of motion identical to the one of the main faults (i.e. they are *synthetic* to the M-surface) and are very often striated (Figs. 2.13 and 2.14). Tension gashes intersect the M-surface at comparatively steeper angles of 30°–50° (Figs. 2.10 and 2.13). They are opening fractures (mode I), commonly filled with minerals. Both R-shears and

FIG. 2.11 (A) Mechanism of formation of mineral steps in protected cavities (i.e. releasing bends) along fault surfaces; and (B) example of mineral (calcite in the present case) steps found on a loose block of Devonian limestone (Xivares Beach, Asturias, Spain); the arrow shows the sense of motion of the missing block, as the block is placed, the sense of slip is sinistral strike slip.

T-fractures make acute angles with the M-surface, these angles opening towards the direction opposite to the sense of motion of the block hosting them (Fig. 2.13).

Two other kinds of secondary fractures were evidenced in the course of laboratory experiments, i.e. *R′-shears* and *P-shears* (Fig. 2.13). These are however rather difficult to identify in the field in general. R′-shears are the conjugate (see Chapter 4 Section 4.4.1 for formal definition) of R-shears and are, therefore, *antithetic* to the M-surface (i.e. display opposite sense of motion). P-shears develop as synthetic shears at very small angles from the M-surface but in a (quasi)symmetric position with respect to R-shears.

FIG. 2.12 (A) Mechanism of formation of stylolitic peaks in restraining bends along fault surfaces; and (B) example of stylolitic peaks along a dextral fault (with minor reverse component) affecting Upper Cretaceous limestones (Amadorio, Alicante, Spain); the arrow shows the sense of motion of the missing block.

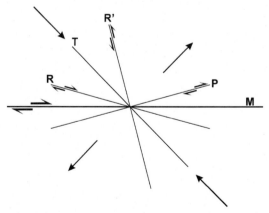

FIG. 2.13 The distinct types of secondary fractures associated with a fault (i.e. M-surface): R=R-shear, R′=R′-shear, P=P-shear and T=T-fracture (tension). Converging and diverging arrows depict directions of shortening and dilatation, respectively.

FIG. 2.14 Example of R-shears along a dextral fault affecting Late Proterozoic (Cryogenian) meta-sandstones (Finnmark, Norway). Note (1) small angles separating the R-shears from the surface of the main fault (M-surface) and (2) intersections between M-surface and R-shears perpendicular to slickenlines. The arrow shows the sense of motion of the missing block.

2.4 Chronology criteria

2.4.1 Stratigraphic control

The timing of fault slip or, in more general terms, of brittle deformation can be controlled by existing relationships between structure and stratigraphy. The age of the rocks affected by the fractures gives a lower limit to the timing of these fractures. Fractures sealed by sediments or lava flows are of particular importance in furnishing upper limits to their ages. For example, fractures measured in Jurassic rocks but absent in Miocene formations were formed during the time interval between Jurassic and Miocene. Thorough field mapping has the potential of revealing the ages of the measured fracture populations, and therefore of the associated tectonic events, with a degree of accuracy that depends strongly on exposure conditions.

Synsedimentary deformation (i.e. deformation of sediments while being deposited) is the most direct stratigraphic criterion for determining the absolute timing of fault slip. Note that synsedimentary deformation is in general easier to detect for normal faults than for reverse or strike-slip faults. In the case of normal faults, increase in sediment thickness consecutive to increase in accommodation space above the hanging wall is particularly demonstrative of the timing of faulting.

2.4.2 Relative chronologies

Where part of the stratigraphic succession is missing, it is difficult to control the respective ages of successive phases of brittle deformation. In turn, it is possible to use *cross-cutting relationships*, between the different types of fractures measured in the field, to determine relative chronologies, though not absolute ages, and to sort out the relative timings of the different phases of deformation. Fig. 2.15A gives an example of cross-cutting relationship

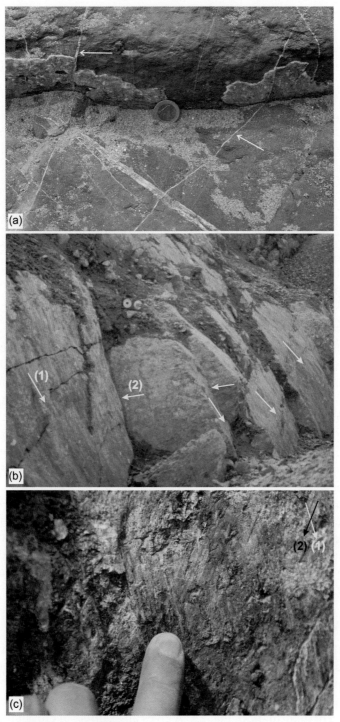

FIG. 2.15 See the figure legend on opposite page

between two sets of quartz veins. The veins shown by the white arrows truncate clearly the other veins and are therefore younger. We should note that one cannot exclude that all the veins belong to a single common tectonic event in the present case. The example of Fig. 2.15B, where a normal fault has been later offset by nearly orthogonal dextral faults, is more informative. As it will be discussed in Chapter 4, the geometrical relationship between the normal fault and the dextral ones calls for a marked change in stress directions between the two faulting events and is suggestive of distinct tectonic events.

Particularly useful in paleostress analyses, though perhaps more demanding for the field geologist, are potential cross-cutting relationships existing between different generations of e.g. striae on the same fault surface. As a rule of thumb, the family of striae that truncates and erases the other ones represents the youngest slip event (Fig. 2.15C).

2.5 Field measurements and writing conventions

Paleostress reconstructions require the preliminary measuring of tectonic structures (mostly fractures) in the field. The gathered information needs to be coded, in particular when mathematical methods and computer programs are used for the final quantification of stress parameters. The studied structures approach, in most cases, either planar or linear geometries and, therefore, two angles are sufficient to describe the attitude of these objects in the three-dimensional space: one angle in the horizontal plane and referred to magnetic north, and a second one in a vertical plane and referred to horizontal or vertical (i.e. to the local gravity vector).

The attitude of a plane is traditionally given as combinations of *strike* and *dip* or *dip direction* and dip (Fig. 2.16A and B), where strike and dip direction are reference angles in the horizontal plane, ranging between 0° and 360° with respect to north, and dip gives the inclination of the plane with respect to horizontal and varies between 0° (horizontal plane) and 90° (vertical plane). In practice, dip direction and dip are measured along the dip line or maximum slope of the plane (i.e. the intersection line between the plane and the vertical plane orthogonal to it), whereas strike is measured parallel to the intersection line between the plane and a horizontal plane. To note, the 'strike and dip' notation must be completed with the information about the (approximate) dip direction (i.e. N, S, W, E or * if the plane is horizontal or vertical).

The attitude of a line (Fig. 2.16C and D) is described in terms of *trend* and *plunge*, where trend is the angle between the projection of the line on the horizontal plane and the

---

FIG. 2.15 (A) Cross-cutting relationships between different generations of quartz veins in Permo-Triassic sandstones; the white arrows indicate the youngest vein set (Xivares Beach, Asturias, Spain); (B) normal fault offset by younger dextral strike-slip faults in Caledonian schists (Mainland Shetland, UK) and (C) dip-slip striae obliterated by younger oblique-slip ones in Triassic limestones (Sichuan, China), note that the oldest generation of striae is particularly well preserved in protected cavities along the fault surface. In (B) and (C), the *arrows* show the sense of motion of the missing blocks; chronological order of faulting events is indicated by the numbers.

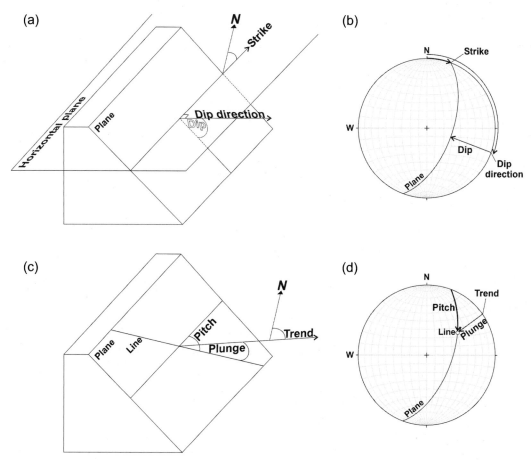

FIG. 2.16 (A), (B) Attitude of a plane in the 3D space as defined by strike and dip or dip direction and dip; (C), (D) attitude of a line in a plane of the 3D space as defined by trend and plunge or pitch. The elements are plotted on the lower hemisphere of a Schmidt net in (B) and (D).

direction of magnetic north, ranging between 0° and 360°, and plunge the angle between the line and the horizontal plane, ranging between 0° and 90°.

We have seen previously (cf. Section 2.3.2) that faulting is prone to create linear features on fault planes (i.e. striae, stylolithes, etc.), which in turn are particularly useful to determine accurately fault slip. The attitude of these linear features may be described in terms of trend and plunge. It is however more convenient to consider that the line is contained in the fault plane, whose attitude is nevertheless determined in the course of the study, and to measure only one reference angle furnishing the attitude of the line within its plane (Fig. 2.16C and D). The latter choice presents the advantage of acquiring precise data while reducing the number of measurements, but in case of shallow-dipping planes, the measurement of trend and plunge is preferred. The angle between the line and a horizontal line in the plane (i.e. a strike line) is obviously the most convenient angle one can measure in the field. Such an angle is referred to as *rake* or *pitch*. Different definitions for both rake

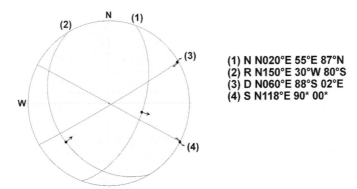

(1) N N020°E 55°E 87°N
(2) R N150°E 30°W 80°S
(3) D N060°E 88°S 02°E
(4) S N118°E 90* 00*

FIG. 2.17 Writing conventions for the full description of fault plane attitudes and related senses of slip. Each sequence represents strike, dip and pitch (see text for details). N = Normal, R = reverse, D = dextral and S = sinistral. The arrows suggest main faulting type (i.e. centrifugal, centripetal and double-paired arrows for normal, reverse and strike-slip faulting, respectively).

and pitch are proposed in the literature and the two words are sometimes used as synonyms; we will describe only the most commonly adopted definitions in the specific seismotectonics and paleostress literature.

Traditionally, rake is defined between −180° and 180°, where negative and positive signs represent normal and reverse fault slip components, respectively. Following the right hand rule, 0° indicates sinistral strike-slip (i.e. the fault block to the right of the observer moves ahead) and ± 180° dextral strike-slip (i.e. the fault block to the right of the observer moves backwards). This latter definition is mainly used by seismologists.

Pitch is mainly used by structural geologists and defined as an angle ranging between 0° (strike-slip) and 90° (dip slip). Note that pitch is measured from the horizontal line in the plane downwards. As such the information needs to be supplemented with the approximate direction the acute angle opens to (i.e. N, S, E or W), except when the pitch equals 0° or 90° (then the sign * follows the angle), and the main component of fault slip (i.e. normal, reverse, dextral or sinistral).

To conclude, many different notations can be introduced to represent field data, although all of them stem from the same geometrical concepts and are gathered using the same type of instruments, i.e. magnetic compass and clinometer. Otherwise stated, we will use in this book the following classical conventions. The attitude of a plane will be characterised by strike and dip (plus approximate dip direction). For example, a plane parallel to the strike line 020–200 and dipping 55° to the East will be written: N020°E 55°E. By definition, three and two digits are needed to fully quantify strike and dip, respectively. Note that the letters 'N' and 'E' encompassing the strike value recall that strike is measured clockwise from north (i.e. 'turning to east'). In the case of a striated fault plane, pitch (two digits) and sense of motion are added to the information: N N020°E 55°E 87°N (Fig. 2.17). The latter example informs that the sense of slip is normal (N) and the motion nearly dip slip, the acute angle between the stria and the strike of the plane opening to north. A reverse fault will be noted 'R' or 'I', a dextral strike-slip fault 'D', and a sinistral strike-slip fault 'S'.

Further reading

Angelier, J., 1994. Paleostress analysis of small-scale brittle structures. In: Hancock, P. (Ed.), Continental Deformation. Pergamon Press, pp. 53–100 (Chapter 4).

Doblas, M., 1998. Slickenside kinematic indicators. Tectonophysics 295, 187–197.

Hancock, P.L., 1985. Brittle microtectonics: principles and practice. J. Struct. Geol. 7, 437–457.

Petit, J.P., 1987. Criteria for the sense of movement on fault surfaces in brittle rocks. J. Struct. Geol. 9(Shear Criteria in Rocks), 597–608.

Twiss, R.J., Moores, E.M., 2007. Structural Geology, second ed. W. H. Freeman, New York.

3 ▪▪▪
▪▪▪
▪▪▪

Theoretical aspects

3.1 From force to stress and stress tensor

3.1.1 Brief historical overview

While the concept of force is intimately wedded to the notion of motion, the concept of stress is associated with deformation and strength of materials. The concept of stress developed progressively from the rupture tests on rods under tension conducted by Galileo (1564–1642) in the early 17th century (Timoshenko, 1953). Although the fact was empirically known since the dawn of civilisation, Galileo demonstrated experimentally that the 'absolute resistance to fracture' of rods under tension is proportional to their cross-sectional area, and introduced implicitly the idea that loads (or forces) are distributed on the surfaces they are applied.

The concept of stress was further investigated and refined, and can be attributed to numerous famous European mathematicians and physicists. In particular, Jakob Bernouilli (1654–1705) and Leibniz (1646–1716) assumed internal tensions acting across surfaces inside a deformed body by the end of the 17th century. In his very last paper in 1705, Jakob Bernouilli made a final but notable contribution, when concluding that a rigorous way to assess deformation of a fibre under tension was to consider force per unit area (i.e. stress) as a function of elongation per unit length (i.e. strain). The statement influenced Euler (1707–1783), who later proposed a linear relationship between stress and strain, introducing the modern form of Hooke's law. Euler also introduced the concept of normal compressive stress, assimilating it to a pressure in a fluid, while Coulomb (1736–1806) explored extensively the notion of shear stress, advanced previously by Parent (1666–1726), by means of laboratory experiments. Lastly, a rigorous and complete mathematical formulation of 3D stress was achieved by Cauchy (1789–1857) in 1822, as part of his theory of linear elasticity of isotropic solids.

3.1.2 Definitions of force and stress

The concept of force is intuitive, as we all have pulled objects of variable sizes and weights with the help of a rope or a string. *Force* is mathematically represented by a vector, giving a direction (e.g. the one towards which one pulls the object with the rope) and a magnitude (e.g. the physical effort one needs to furnish to move the object). Two main types of forces can be distinguished depending on their mode of application. Forces applied to the edges of an object or acting across a surface within the object are called *surface forces*. Forces acting on each particle of an object, including the ones within the object, are termed *body forces*. Such forces include gravitational or electromagnetic forces and act 'at distance'

Paleostress Inversion Techniques. https://doi.org/10.1016/B978-0-12-811910-5.00007-5

without direct contact between two bodies (a concept put forward by Newton, taking aback most scientists of the 17th century!). When superposed, surface and body forces result in net forces acting across surface elements of the object.

As Galileo and his followers figured out, in order to understand deformation of solids, the concept of force is in itself insufficient if the area of the surface on which the force acts is not considered. It is thus more convenient to introduce a quantity describing how forces are distributed on the surfaces they act or, conversely, furnishing a measure of force intensity on these surfaces:

$$\vec{\sigma} = \frac{\vec{F}}{S}$$ (3.1)

where $\vec{F}$ is the force, S is the area and $\vec{\sigma}$ is the stress (vector) acting on the surface, also called *surface stress* (to note, the vector $\vec{\sigma}$ is often called 'traction' in traditional mechanics literature; we follow here the common wording used in Earth sciences and other disciplines). By definition, stress is expressed in Newtons per unit area (N/m^2) or Pascals (Pa), as it is commonly done for pressures. Tectonic stresses reach typically tens of MPa.

Eq. (3.1) represents stress in its simplest form and is only valid if the force remains constant in both direction and magnitude regardless of the location on the surface. A more complete and rigorous mathematical definition is derived by expressing stress for an infinitesimally small area containing a given point on the surface, which is equivalent to defining stress at a point on the surface (Fig. 3.1). Physically, an 'infinitesimally small area' has to be regarded as a sufficiently small area for which potential variations of the force vector remain negligible; such a sufficiently small area cannot be reduced down to, e.g., the typical dimensions of the atomic structure of the material because classical mechanics is not applicable at such scales.

In order to derive a more complete definition of stress, let us consider a body containing a small element of surface ΔS separating the body in two parts (Fig. 3.1) or, conversely, two bodies in contact across ΔS. Let $\Delta \vec{F}$ be the elementary surface force exerted by (1) on (2) across ΔS. Thus, we can write

$$\vec{\sigma} = \lim_{S \to 0} \frac{\Delta \vec{F}}{\Delta S}$$ (3.2)

$$\text{or} \quad \vec{\sigma} = \frac{d\vec{F}}{dS}$$ (3.3)

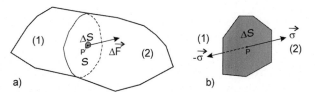

FIG. 3.1 Definition of surface stress. (A) Body (1) exerts an elementary force (2) across an element of surface ΔS. (B) Representation of surface stress at point P.

Eq. (3.3) ensures that the stress vector can always be defined. According to Newton's Third Law (i.e. principle of action and reaction), any applied force is counteracted by a force of equal magnitude but opposite in direction. Thus, in the example given in Fig. 3.1, $\vec{\sigma}$ is counteracted by a stress of equal magnitude exerted by (2) on (1). The practical implication of the latter is that stresses do not furnish any information about the relative position of the 'push' or the 'pull' from which they originate. For instance, an N—S stress determined in the field, either using in situ geophysical measurements or paleostress inversion methods, can result from a load located either north or south of the measurement site. Accordingly, stresses are generally represented by pairs of arrows of equal lengths but pointing in opposite directions.

A surface stress applied inwards on a body has the tendency to shorten the body; thus, it is referred to as *compressive stress*, and by convention, its sign is positive and is graphically represented by two arrows facing one another (Fig. 3.2A). A *tensile stress* implies surface stress pointing outwards from the body has the tendency to stretch the body; it takes a negative value and is represented by diverging arrows (Fig. 3.2B). Note that the adopted sign convention is arbitrary. The convention given here, sometimes referred to as 'geologic tensor sign convention', is the one traditionally used in geological and soil sciences but is opposite to what is normally preferred in continuum mechanics and material science! The choice is motivated by the fact that, owing to the overburden, stresses are generally compressive in the underground.

Let us focus on the physical nature of the stress vector. To these aims, we consider the simple 2D problem depicted in Fig. 3.3, where an infinitesimally small triangular body is submitted to normal forces applied to its lower and left external boundaries. The long edge of the triangle contacts a second body (or can also be taken as one plane within a larger body, whose upper and right boundaries are not represented here), which in turn resists the applied external forces according to the principle of action-reaction:

$$\vec{F}_S = -\left(\vec{F}_x + \vec{F}_y\right) \tag{3.4}$$

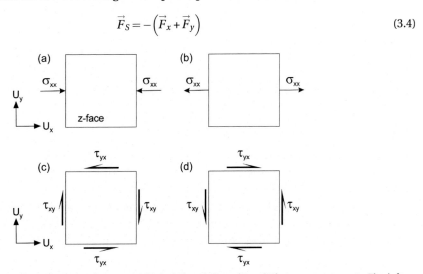

FIG. 3.2 'Geologic tensor sign convention' for normal (A and B) and shear (C and D) stress components. The left column shows the stresses noted positive, and the right column shows the stresses noted negative.

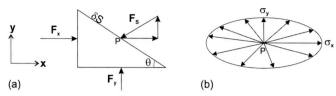

FIG. 3.3 (A) Force balance condition for a triangular element loaded across its left and lower boundaries; (B) representation of the collection of permissible stress vectors at P as a function of the attitude (i.e. angle θ) of δS.

where $\vec{F}_S$ is the reaction force on the surface δS.

Considering only absolute values, we derive from Eq. (3.4):

$$F_{Sx} = F_x \tag{3.5}$$

$$F_{Sy} = F_y \tag{3.6}$$

Expressing forces in function of stresses, Eqs (3.5) and (3.6) become

$$\sigma_{Sx}\partial S = \sigma_x \partial S \sin\theta \tag{3.7}$$

$$\sigma_{Sy}\partial S = \sigma_y \partial S \cos\theta \tag{3.8}$$

and finally

$$\sigma_{Sx} = \sigma_x \sin\theta \tag{3.9}$$

$$\sigma_{Sy} = \sigma_y \cos\theta \tag{3.10}$$

Note that the development shows that the components of the stress vector depend on the angle θ.

Re-arranging and squaring Eqs (3.9) and (3.10), we get

$$\left(\frac{\sigma_{Sx}}{\sigma_x}\right)^2 = \sin^2\theta \tag{3.11}$$

$$\left(\frac{\sigma_{Sy}}{\sigma_y}\right)^2 = \cos^2\theta \tag{3.12}$$

Let us add Eq. (3.11) to Eq. (3.12):

$$\left(\frac{\sigma_{Sx}}{\sigma_x}\right)^2 + \left(\frac{\sigma_{Sy}}{\sigma_y}\right)^2 = 1 \tag{3.13}$$

We obtain the characteristic equation of an ellipse, whose semi-major (long) and semi-minor (short) axes are respectively σ_x and σ_y in the present case. In other words: *(1) the stress vector varies in orientation and magnitude when the attitude of the surface δS, i.e. the angle θ, changes; (2) its magnitude is constrained to remain between σ_y (minimum stress in the present case) and σ_x (maximum stress); (3) the whole collection of physically possible stress vectors forms an ellipse, the stress ellipse or Lamé's stress ellipse.*

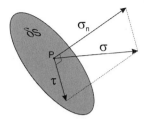

FIG. 3.4 Definition of normal ($\vec{\sigma}_n$) and shear ($\vec{\tau}$) stress vectors.

A convenient way to manipulate a stress vector is to break it down into two components: one component perpendicular to the surface, the *normal stress* vector $\vec{\sigma}_n$, and the other parallel to it, the *shear stress* vector $\vec{\tau}$ (Fig. 3.4):

$$\vec{\sigma} = \vec{\sigma}_n + \vec{\tau} \tag{3.14}$$

We will see in the following text that the choice is particularly elegant when dealing with graphical representation of stresses acting on a plane and theoretical laws for brittle deformation.

3.1.3 Definition of stress tensor

However, the knowledge of surface stress, $\vec{\sigma}$, is limited to the specific surface under consideration as $\vec{\sigma}$ is dependent on the surface orientation. Ideally, one would like to use a mathematical tool that would characterise the stress at a point P in the 3D space regardless of the surface orientation. One such tool is the *stress tensor* introduced originally by Cauchy, often named *Cauchy stress tensor*.

In order to define the stress tensor, T_0, it is convenient to consider the stresses applied on three mutually perpendicular faces of an infinitesimally small cube centred on P (Fig. 3.5). That is, the cube is small enough to assume that the stresses act directly on P. In virtue of Newton's Third Law, the stress on each of the three concealed faces is symmetrical to its respective counterpart on the parallel face. Furthermore, it is convenient to choose a coordinate Cartesian system such that each of the three coordinate vectors is perpendicular to one of the three considered faces. The three stress vectors, $\vec{\sigma}_x$, $\vec{\sigma}_y$ and $\vec{\sigma}_z$, fully define the stress at P, which means that the stress vector on any surface containing P can be derived from these three independent vectors by means of mathematical transformation (i.e. *Cauchy's fundamental theorem*). Conventionally, the subscript of each one of the three independent stress vectors refers to the face it is attached to or conversely to the outward direction of its normal.

We have seen that a practical way to treat stress vectors consists of expressing them as a function of their respective normal and shear components (Fig. 3.5):

$$\vec{\sigma}_x = \sigma_{xx}\vec{U}_x + \tau_{xy}\vec{U}_y + \tau_{xz}\vec{U}_z \tag{3.15}$$

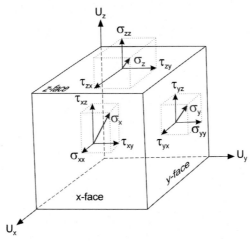

FIG. 3.5 The 'Cauchy stress cube' or 3D representation of all stress components acting on an infinitesimal volume. See text for explanation of symbols.

$$\vec{\sigma}_y = \tau_{yx}\vec{U}_x + \sigma_{yy}\vec{U}_y + \tau_{yz}\vec{U}_z \tag{3.16}$$

$$\vec{\sigma}_z = \tau_{zx}\vec{U}_x + \tau_{zy}\vec{U}_y + \sigma_{zz}\vec{U}_z \tag{3.17}$$

The first subscript indicates the face on which the stress component acts (i.e. the coordinate direction of its outward normal), and the second subscript indicates the coordinate direction in which the stress component acts. The latter convention is called the 'on-in convention of subscripts'.

The three linear Eqs (3.15), (3.16) and (3.17) can be written in matrix form:

$$\begin{pmatrix} \sigma_x \\ \sigma_y \\ \sigma_z \end{pmatrix} = \begin{pmatrix} \sigma_{xx} & \tau_{xy} & \tau_{xz} \\ \tau_{yx} & \sigma_{yy} & \tau_{yz} \\ \tau_{zx} & \tau_{zy} & \sigma_{zz} \end{pmatrix} \begin{pmatrix} U_x \\ U_y \\ U_z \end{pmatrix} \tag{3.18}$$

where the stress tensor as expressed in the chosen Cartesian reference system is

$$T_0 = \begin{pmatrix} \sigma_{xx} & \tau_{xy} & \tau_{xz} \\ \tau_{yx} & \sigma_{yy} & \tau_{yz} \\ \tau_{zx} & \tau_{zy} & \sigma_{zz} \end{pmatrix} \tag{3.19}$$

The stress components are now grouped in an array matrix, which is a second-rank tensor, where each row contains the stress components acting on one face of the cube and each column describes the stress components acting according to a particular coordinate direction. The nine components of the stress tensor describe all possible stress vectors acting on their corresponding planes at P, i.e. they describe the *stress state* at point P. Eq. (3.18) defines the stress tensor as a linear operator, whose application on a unit vector of the 3D space furnishes the components of the stress vector attached to the plane perpendicular to this unit vector (i.e. a triangular planar element slicing the cube in the present case; see Section 3.1.4).

The assumption of static stress state implies not only force balance but also moment balance or, in other words, no rotation of the cube. The latter statement translates into a set of conditions where opposite torques arising from the shear components counteract one another:

$$\tau_{ij} = \tau_{ji} \qquad (3.20)$$

with i and j being successively x, y and z and $i \neq j$.

The number of independent components is then reduced to six, and the stress tensor shows the important property of being symmetrical:

$$T_0 = \begin{pmatrix} \sigma_{xx} & \tau_{xy} & \tau_{xz} \\ \tau_{xy} & \sigma_{yy} & \tau_{yz} \\ \tau_{xz} & \tau_{yz} & \sigma_{zz} \end{pmatrix} \qquad (3.21)$$

Before proceeding with further properties of the stress tensor, we need to set up sign conventions for its non-diagonal shear components. As for the diagonal normal components we use the 'geologic tensor sign convention'. Positive shear stresses act in positive (negative) coordinate directions on cube edges facing negative (positive) directions (Fig. 3.2C). Shear stresses acting in positive (negative) coordinate directions on cube edges facing positive (negative) directions are negative (Fig. 3.2D). The reader should note that this convention is not universally adopted, even within the tectonics and structural geology community (cf. the extensive discussion in Twiss and Moores, 2001, p. 187). It is strongly recommended to identify what specific convention is used in the literature in order to avoid potential confusion.

It is always possible by means of careful rotations of the cube to find the unique configuration for which all three stress vectors become strictly perpendicular to the faces they are applied or, conversely, for which all shear stress components vanish. This operation is equivalent to determining mathematically the three eigenvectors and the three eigenvalues of the stress tensor. By definition, the eigenvectors of a matrix form an orthogonal base.

The components of the stress tensor are consequently rewritten in the reference system of its eigenvectors $(\vec{U}_1, \vec{U}_2, \vec{U}_3)$:

$$T = \begin{pmatrix} \sigma_1 & 0 & 0 \\ 0 & \sigma_2 & 0 \\ 0 & 0 & \sigma_3 \end{pmatrix} \qquad (3.22)$$

$$\text{with} \quad \vec{\sigma}_1 = \sigma_1 \vec{U}_1 \qquad (3.23)$$

$$\vec{\sigma}_2 = \sigma_2 \vec{U}_2 \qquad (3.24)$$

$$\vec{\sigma}_3 = \sigma_3 \vec{U}_3 \qquad (3.25)$$

The eigenvalues of the stress tensor, or magnitudes of the *principal stresses*, are σ_1 (i.e. *maximum principal stress*), σ_2 (i.e. *intermediate principal stress*) and σ_3 (i.e. *minimum principal stress*). The three eigenvectors give the directions of the three principal stresses. Pairs of eigenvectors define the *principal planes of stress*.

As common in Earth Sciences, we write

$$\sigma_1 \geq \sigma_2 \geq \sigma_3 \tag{3.26}$$

As stated before, underground stresses are generally compressive (i.e. positive), but the relationship (Eq. (3.26)) also involves the cases where principal stresses are tensile (i.e. negative). For example, if the three principal stresses are negative, σ_1 represents the minimum tensile stress and σ_3 the maximum tensile stress. Examining Eq. (3.26) carefully, we introduce some special stress states:

(1) if $\sigma_1 = \sigma_2 = \sigma_3 = p$, the stress state is *hydrostatic* (i.e. isotropic) and equivalent to the applied pressure
(2) if $\sigma_1 > \sigma_2 = \sigma_3$ or $\sigma_1 = \sigma_2 > \sigma_3$, the stress state is *biaxial*
(3) if $\sigma_1 > 0$ and $\sigma_2 = \sigma_3 = 0$ or $\sigma_3 < 0$ and $\sigma_1 = \sigma_2 = 0$, the stress state is *uniaxial*, which is a particular case of the biaxial stress state

These definitions are traditionally used in the fields of engineering and rock mechanics. However, it is very common in geological literature to name 'uniaxial' or '*radial*' the two stress states listed under (2), with the background idea that one principal stress direction 'rules' the system, deformation being isotropic according to the perpendicular principal stress plane. Unless stated otherwise, we will use the geological wording.

The stress tensor, T, is a mathematical object much easier to handle than its counterpart, T_0, as only three components are now displayed in the diagonal form. The latter formulation of the stress tensor can be somewhat confusing. Three components are explicitly given in Eq. (3.22); however, we have previously found that the stress tensor is defined by six components (see Eq. (3.21)). The three components, which are 'missing', are the three angles defining the respective directions of the three eigenvectors that, in turn, are not explicitly expressed in the written form of the tensor, T. One has to keep in mind that T is expressed in the reference frame of its eigenvectors, themselves referred to some convenient reference frame in the physical world (hence the three angles), e.g. the geographical system of coordinates supplemented with the vertical direction.

Let us consider again the respective expressions of T_0 (Eq. (3.21)) and T (Eq. (3.22)). As a consequence of the change in the coordinate system, the values of the scalar components differ from one expression of the stress tensor to the other. However, one expects T_0 and T to describe an identical physical condition, which by definition shall not be dependent on an arbitrary choice of the coordinate system. Conservation of the physics of the system is ensured by the three *stress invariants*, given in the $(\vec{U}_x, \vec{U}_y, \vec{U}_z)$ coordinate system by

$$I_1 = \sigma_{xx} + \sigma_{yy} + \sigma_{zz} \tag{3.27}$$

$$I_2 = \sigma_{xx}\,\sigma_{yy} + \sigma_{yy}\,\sigma_{zz} + \sigma_{zz}\,\sigma_{xx} - \tau^2_{xy} - \tau^2_{yz} - \tau^2_{zx} \tag{3.28}$$

$$I_3 = \sigma_{xx}\,\sigma_{yy}\,\sigma_{zz} + 2\tau_{xy}\,\tau_{yz}\,\tau_{zx} - \sigma_{xx}\,\tau^2_{yz} - \sigma_{yy}\,\tau^2_{zx} - \sigma_{zz}\,\tau^2_{xy} \tag{3.29}$$

and in the $(\vec{U}_1, \vec{U}_2, \vec{U}_3)$ coordinate system (wherein shear components disappear) given by

$$I_1 = \sigma_1 + \sigma_2 + \sigma_3 \tag{3.30}$$

$$I_2 = \sigma_1 \sigma_2 + \sigma_2 \sigma_3 + \sigma_3 \sigma_1 \tag{3.31}$$

$$I_3 = \sigma_1 \sigma_2 \sigma_3 \tag{3.32}$$

Another important property of the stress tensor is that it can always be expressed as the sum of a deviatoric component and a hydrostatic component:

$$T = T_D + pI \tag{3.33}$$

where T_D is the *stress deviator*, a special stress tensor describing a state of pure shear; I is the unit matrix; and p is a positive scalar, the *mean stress*, which is an isotropic pressure such that

$$p = \frac{1}{3}(\sigma_1 + \sigma_2 + \sigma_3) \tag{3.34}$$

$$\text{or } p = \frac{1}{3}I_1 \tag{3.35}$$

Eq. (3.33) can be rewritten as

$$\begin{pmatrix} \sigma_1 & 0 & 0 \\ 0 & \sigma_2 & 0 \\ 0 & 0 & \sigma_3 \end{pmatrix} = \begin{pmatrix} \sigma_1 - p & 0 & 0 \\ 0 & \sigma_2 - p & 0 \\ 0 & 0 & \sigma_3 - p \end{pmatrix} + p \begin{pmatrix} 1 & 0 & 0 \\ 0 & 1 & 0 \\ 0 & 0 & 1 \end{pmatrix} \tag{3.36}$$

$$\text{and } T_D = \frac{1}{3} \begin{pmatrix} 2\sigma_1 - \sigma_2 - \sigma_3 & 0 & 0 \\ 0 & 2\sigma_2 - \sigma_3 - \sigma_1 & 0 \\ 0 & 0 & 2\sigma_3 - \sigma_1 - \sigma_2 \end{pmatrix} \tag{3.37}$$

We note that the sum of the diagonal terms of T_D, its first invariant, I_1, is equal to zero, which is a general property of stress deviators.

Finally, we introduce an ellipsoid whose semi-axes in the stress space are $\vec{\sigma}_1$, $\vec{\sigma}_2$ and $\vec{\sigma}_3$. This ellipsoid is called the stress ellipsoid, and it represents the natural 3D extension of the stress ellipse we defined previously (Fig. 3.3). The orientation of the stress ellipsoid is dictated by the orientation of the principal stresses, and its shape depends on the relative lengths, or magnitudes, of $\vec{\sigma}_1$, $\vec{\sigma}_2$ and $\vec{\sigma}_3$.

In the particular case of hydrostatic stress, the stress ellipsoid adopts the geometry of a sphere, whose radius is equal to the hydrostatic pressure. In the general case, and in order to quantify the shape of the stress ellipsoid, it is convenient to introduce a dimensionless quantity based on a ratio between principal stress magnitudes differences, namely the *shape ratio* or *stress ratio*:

$$\Phi = \frac{\sigma_2 - \sigma_3}{\sigma_1 - \sigma_3} \tag{3.38}$$

$$\text{with } 0 \leq \Phi \leq 1 \tag{3.39}$$

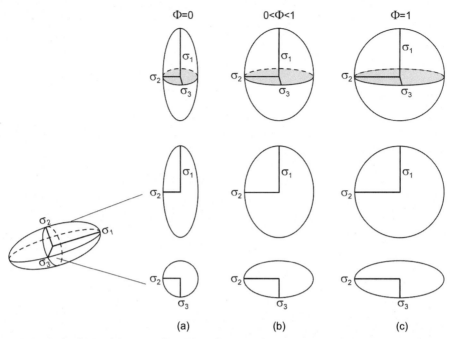

FIG. 3.6 Variation in the shape of the stress ellipsoid as a function of the shape ratio, Φ. (A) Prolate shape for $\Phi = 0$ (i.e. $\sigma_2 = \sigma_3$). (B) Intermediate shape for $0 < \Phi < 1$ (i.e. $\sigma_3 < \sigma_2 < \sigma_1$). (C) Oblate shape for $\Phi = 1$ (i.e. $\sigma_2 = \sigma_1$).

The ratio Φ, sometimes called R, is the most commonly used ratio in paleostress studies. However, one should note that other definitions of the shape ratio, albeit with the same meaning, are often found in the literature. The ratio Φ varies from 0 (when $\sigma_2 = \sigma_3$) to 1 (when $\sigma_2 = \sigma_1$); thus, it maps all possible stress states, pertaining to a particular orientation of the stress axes, from one biaxial stress state (or 'uniaxial' according to the geological wording) to the other. In other words, the ratio Φ represents the degree of anisotropy of the stress state. Each value of Φ corresponds to a specific shape of the stress ellipsoid that, in turn, characterises a precise stress state. Accordingly, for $\Phi = 0$, the stress ellipsoid takes a cigar-like geometry (i.e. prolate shape), and for $\Phi = 1$, its shape resembles that of a pancake (i.e. oblate shape); in between the two values, the stress ellipsoid adopts intermediate forms (Fig. 3.6).

3.1.4 Stress on planes

The determination of stresses acting on planes is the backbone of paleostress studies, particularly but not exclusively, when these rely on fault slip inversion (Chapter 4) or calcite twin data inversion (Chapter 6). Thus, it is essential to examine carefully some fundamental aspects before introducing paleostress inversion methods in the following chapters.

One can derive a set of useful equations in a relatively easy way by adopting the classical approach introduced by Cauchy and considering a plane slicing the 'stress cube'

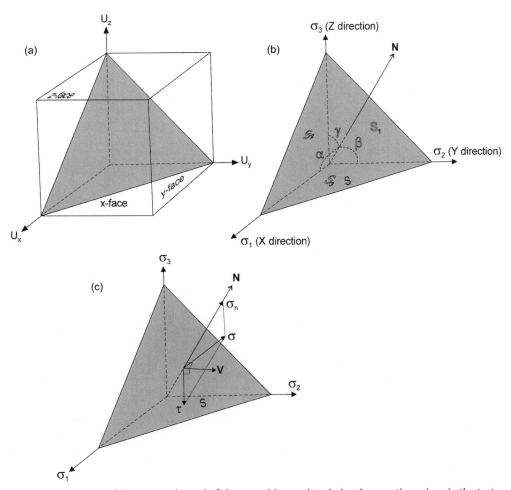

FIG. 3.7 Representation of the stress cube and of the quantities used to derive the equations given in the text.

(Fig. 3.5) as shown in Fig. 3.7. Furthermore, it is convenient to assume that the principal stress axes are parallel to the reference axes (i.e. we are now working in the stress space; see Section 3.1.3). The assumption results in a drastic, though welcome, simplification of the mathematical analysis. Nevertheless, the physical properties of the system are conserved and can be shown and discussed based on the simple equations that are developed hereafter.

The orientation of the normal of the plane, $\vec{N}$, is defined by the three angles α, β and γ, or by its three direction cosines(Fig. 3.7):

$$l = \cos(\alpha) \tag{3.40}$$

$$m = \cos(\beta) \tag{3.41}$$

$$n = \cos(\gamma) \tag{3.42}$$

The normal vector, $\vec{N}$, is a unit vector; thus

$$l^2 + m^2 + n^2 = 1 \tag{3.43}$$

Let us determine the stress vector, $\vec{\sigma}$, acting on the plane whose surface area is S. A natural first step is to write the stress vector as a function of its X, Y and Z components:

$$\vec{\sigma} = \begin{pmatrix} \sigma_x \\ \sigma_y \\ \sigma_z \end{pmatrix} \tag{3.44}$$

We now search for the expression of each component as a function of σ_1, σ_2 and σ_3, following the same line of reasoning as in Fig. 3.3 but extended to the third dimension. We name the four planes according to their respective surface areas, S, S_1, S_2 and S_3. Force balance implies that, for example, the force parallel to the X direction and acting on plane S is counteracted by a force of the same magnitude parallel to the X direction (i.e. parallel to σ_1) and acting on plane S_1. Considering only the absolute force values, the latter statement translates to

$$S\sigma_X = S_1\,\sigma_1 \tag{3.45}$$

We note that plane S_1 is the orthogonal projection of plane S according to a rotation of angle α; thus

$$S_1 = \cos(\alpha)\,S \tag{3.46}$$

$$\text{or } S_1 = lS \tag{3.47}$$

Eq. (3.45) can be rewritten as

$$\sigma_X = l\sigma_1 \tag{3.48}$$

$$\text{Similarly}: \quad \sigma_Y = m\sigma_2 \tag{3.49}$$

$$\text{and } \sigma_Z = n\sigma_3 \tag{3.50}$$

Squarring and adding Eqs (3.48)–(3.50), we note that the magnitude of the stress vector can be expressed as a function of the direction cosines of plane S:

$$\sigma^2 = l^2\,\sigma_1{}^2 + m^2\,\sigma_2{}^2 + n^2\,\sigma_3{}^2 \tag{3.51}$$

We further remark that after squaring, re-arranging and adding Eqs (3.48)–(3.50) and paying attention to the condition given by Eq. (3.43), we get

$$\frac{\sigma_X^2}{\sigma_1^2} + \frac{\sigma_Y^2}{\sigma_2^2} + \frac{\sigma_Z^2}{\sigma_3^2} = 1 \tag{3.52}$$

which is the equation of the stress ellipsoid, whose existence has previously been inferred based on geometrical arguments.

The normal stress vector, $\vec{\sigma}_n$, is the projection of the stress vector, $\vec{\sigma}$, on the normal of the plane S, $\vec{N}$; thus, its magnitude is given by

$$\sigma_n = \vec{\sigma}\vec{N} \tag{3.53}$$

$$\sigma_n = \begin{pmatrix} l\,\sigma_1 \\ m\,\sigma_2 \\ n\,\sigma_3 \end{pmatrix} \begin{pmatrix} l \\ m \\ n \end{pmatrix} \tag{3.54}$$

$$\sigma_n = l^2\,\sigma_1 + m^2\,\sigma_2 + n^2\,\sigma_3 \tag{3.55}$$

Eq. (3.55) demonstrates mathematically that the magnitude of the normal stress component depends solely on the orientation of the plane with respect to the principal stresses. We now determine the extreme values of σ_n.

Let us rewrite Eq. (3.43):

$$n^2 = 1 - l^2 - m^2 \tag{3.56}$$

and combine the latter expression with Eq. (3.55):

$$\sigma_n = l^2\,(\sigma_1 - \sigma_3) + m^2\,(\sigma_2 - \sigma_3) + \sigma_3 \tag{3.57}$$

$$\text{or } \sigma_n = \cos^2(\alpha)\,(\sigma_1 - \sigma_3) + \cos^2(\beta)\,(\sigma_2 - \sigma_3) + \sigma_3 \tag{3.58}$$

To determine the extreme values, we differentiate Eq. (3.58) with respect to α and β:

$$\frac{d\sigma_n}{d\alpha} = -2\,(\sigma_1 - \sigma_3)\,\cos(\alpha)\sin(\alpha) \tag{3.59}$$

$$\frac{d\sigma_n}{d\beta} = -2\,(\sigma_2 - \sigma_3)\,\cos(\beta)\sin(\beta) \tag{3.60}$$

and we search for the values for which the two obtained functions are equal to zero.

After examining Fig. 3.7B, we exclude the case $\alpha = \beta = 0°$, which is geometrically impossible. Therefore, we identify only three distinct cases for which both Eqs (3.59) and (3.60) are equal to zero simultaneously:

$$\alpha = 90°\text{and}\,\beta = 90°, \text{i.e.}\sigma_n = \sigma_3$$

$$\alpha = 90°\text{and}\,\beta = 0°, \text{i.e.}\sigma_n = \sigma_2$$

$$\alpha = 0°\text{and}\,\beta = 90°, \text{i.e.}\sigma_n = \sigma_1$$

These three cases correspond to specific geometrical situations where the plane S is parallel to one of the principal stress planes: S is parallel to the $(\vec{\sigma}_1, \vec{\sigma}_2)$ plane for case (1); to the $(\vec{\sigma}_1, \vec{\sigma}_3)$ plane for case (2); and to the $(\vec{\sigma}_2, \vec{\sigma}_3)$ plane for case (3). Moreover, cases (3) and (1) can be viewed as two particular orientations of $\vec{\sigma}_n$ when this vector rotates in the $(\vec{\sigma}_1, \vec{\sigma}_3)$ plane according to α, and its magnitude increases from σ_3 to σ_1.

In brief, Eq. (3.55) demonstrates two important properties: *(1) the magnitude of the normal stress vector depends on the orientation of the plane with respect to the principal stress axes; (2) the magnitude of the normal stress vector always remains in the interval between σ_3 and σ_1 or, conversely, these two latter values represent the respective minimum and maximum normal stresses that can affect the plane.*

We now focus on the shear stress component. The magnitude of the shear stress vector can be derived by squaring and re-arranging Eq. (3.14):

$$\tau^2 = \sigma^2 - \sigma_n{}^2 \tag{3.61}$$

Combining (3.51), (3.55) and (3.61), we get

$$\tau^2 = l\sigma_1{}^2 + m\sigma_2{}^2 + n\sigma_3{}^2 - \left(l^2\,\sigma_1 + m^2\,\sigma_2 + n^2\,\sigma_3\right)^2 \tag{3.62}$$

Finally, after modifying Eq. (3.62) and combining it with Eq. (3.43), we get

$$\tau^2 = l^2\,m^2\,(\sigma_1 - \sigma_2)^2 + m^2\,n^2\,(\sigma_2 - \sigma_3)^2 + l^2\,n^2\,(\sigma_1 - \sigma_3)^2 \tag{3.63}$$

Let us examine Eq. (3.63). First, we remark that shear stress magnitude depends solely on plane orientation and stress differences. In the case of hydrostatic (i.e. isotropic) stress states, where $\sigma_1 = \sigma_2 = \sigma_3$, $\tau = 0$, i.e. there is no shear stress on the plane. If two of the directions cosines are equal to zero, the third one automatically equals 1 (i.e. two of the three reference angles are equal to 90° and the remaining one is equal to 0°; Fig. 3.7B). These three cases correspond once again to the three geometrical configurations where the plane is parallel to one principal stress plane, i.e. the plane contains two principal stress axes and is perpendicular to the third one. Eq. (3.63) shows that for these configurations, shear stresses vanish again. In conclusion, *a shear stress component exists on a given plane only if the stress state is non-hydrostatic (i.e. anisotropic) and at least two principal stress axes are oblique to the plane.*

After these fundamental considerations on shear stress magnitude, it is particularly useful to discuss also the orientation of the shear stress vector. To these aims, we first write the vector product:

$$\vec{V} = \vec{N} \times \vec{\sigma} \tag{3.64}$$

and according to Eqs (3.40)–(3.42) and (3.48)–(3.50):

$$\vec{V} = \begin{pmatrix} l \\ m \\ n \end{pmatrix} \times \begin{pmatrix} l\,\sigma_1 \\ m\,\sigma_2 \\ n\,\sigma_3 \end{pmatrix} \tag{3.65}$$

$$\vec{V} = \begin{pmatrix} -mn\,(\sigma_2 - \sigma_3) \\ \ln\,(\sigma_1 - \sigma_3) \\ -lm\,(\sigma_1 - \sigma_2) \end{pmatrix} \tag{3.66}$$

We see in Fig. 3.7C that $\vec{V}$ is a vector in the plane S and perpendicular to $\vec{\tau}$ such that counter-clockwise rotation (i.e. in agreement with the orientation of $\vec{N}$) brings $\vec{V}$ parallel to $\vec{\tau}$ and pointing to the same direction.

Hence:

$$\vec{\tau} = \vec{V} \times \vec{N} \tag{3.67}$$

$$\vec{\tau} = \begin{pmatrix} -mn\,(\sigma_2 - \sigma_3) \\ ln\,(\sigma_1 - \sigma_3) \\ -lm\,(\sigma_1 - \sigma_2) \end{pmatrix} \times \begin{pmatrix} l \\ m \\ n \end{pmatrix} \tag{3.68}$$

$$\vec{\tau} = \begin{pmatrix} ln^2\,(\sigma_1 - \sigma_3) + lm^2\,(\sigma_1 - \sigma_2) \\ -l^2 m\,(\sigma_1 - \sigma_2) + mn^2\,(\sigma_2 - \sigma_3) \\ -m^2 n\,(\sigma_2 - \sigma_3) - l^2 n\,(\sigma_1 - \sigma_3) \end{pmatrix} \tag{3.69}$$

Dividing all the components of $\vec{\tau}$ by the *differential stress*, $\sigma_1 - \sigma_3$, and combining the results with the expression of the shape ratio given by Eq. (3.38), we finally obtain

$$\vec{\tau} = (\sigma_1 - \sigma_3) \begin{pmatrix} l\,(n^2 + m^2\,(1 - \Phi)) \\ m\,(n^2 \Phi - l^2\,(1 - \Phi)) \\ -n\,(m^2 \Phi + l^2) \end{pmatrix} \tag{3.70}$$

Eq. (3.70) demonstrates that the orientation of the shear stress vector depends not only on the orientation of the plane with respect to the principal stress axes but also on the value of the shape ratio (but not on the stress differences!). In brief, *the orientation of the shear stress is sensitive to the degree of anisotropy of the stress state*, a perhaps counter-intuitive but fundamental result, which is unfortunately often overlooked.

We also note that, because shear stress components are solely expressed in terms (of ratio) of their differences in Eq. (3.70), the orientation of $\vec{\tau}$ does not change if an isotropic stress is added or subtracted. The latter finding means, in physical terms, that the orientation of the shear stress is insensitive to the addition/subtraction of a pressure. We will see in Chapter 4 that this stress property is essential in paleostress inversion methods involving fault slip data.

3.1.5 Mohr diagrams

We have demonstrated previously that the stress vector, and consequently its normal and shear components, vary with the attitude of the plane (see Eqs (3.55) and (3.63)). In addition, we have shown that the stress state at a given point can be represented graphically by an ellipse in 2D or an ellipsoid in 3D. However, it is particularly tedious to extract normal and shear stress components from such representations. To these aims, the German engineer, Otto Mohr (1835–1918), pursuing the original work of Karl Culmann (1821–1881), devised a very practical graphical tool: the *Mohr diagram* (Mohr 1882). The Mohr diagram consists of a representation of the stress state in the (σ_n, τ) 2D space (i.e. *Mohr space*) in the form of a circle, the *Mohr circle*, and allows for direct reading of $\vec{\sigma}_n$ and $\vec{\tau}$ magnitudes for all planes of the physical space.

In order to derive the mathematical equation of the Mohr circle, we consider once again the problem of force balance under static equilibrium conditions in 2D as depicted in Fig. 3.8.

We first express the normal stress (σ_n) and the shear stress (τ) acting on the long edge of the triangle as a function of the principal stresses.

Force balance according to the orientation of $\vec{\sigma}_1$ is given by

$$\sigma_1(\partial S \cos\theta) = (\sigma_n \cos\theta)\partial S + (\tau \sin\theta)\partial S \tag{3.71}$$

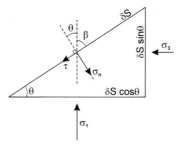

FIG. 3.8 Triangular element loaded across its right and lower edges and representation of reaction stresses along its long edge. See text for explanation of symbols.

and according to the orientation of $\vec{\sigma}_3$ by

$$\sigma_3(\partial S \sin\theta) = (\sigma_n \sin\theta)\partial S - (\tau\cos\theta)\partial S \tag{3.72}$$

Eqs (3.71) and (3.72) are rewritten as follows:

$$\sigma_1 \cos\theta = \sigma_n \cos\theta + \tau\sin\theta \tag{3.73}$$

$$\sigma_3 \sin\theta = \sigma_n \sin\theta - \tau\cos\theta \tag{3.74}$$

We now multiply Eq. (3.73) by $\cos\theta$ and Eq. (3.74) by $\sin\theta$. Adding the two newly obtained equations, we get

$$\sigma_1 \cos^2\theta + \sigma_3 \sin^2\theta = \sigma_n \tag{3.75}$$

Recalling the two following trigonometric identities (i.e. double-angle formulae):

$$\cos^2\theta = \frac{1 + \cos 2\theta}{2} \tag{3.76}$$

$$\sin^2\theta = \frac{1 - \cos 2\theta}{2} \tag{3.77}$$

And combining with Eq. (3.75), we finally get

$$\sigma_n = \frac{\sigma_1 + \sigma_3}{2} + \frac{\sigma_1 - \sigma_3}{2}\cos 2\theta \tag{3.78}$$

We proceed similarly to derive the expression of the shear stress, τ. Let us multiply Eq. (3.73) by $\sin\theta$ and Eq. (3.74) by $-\cos\theta$, and add the two obtained equations:

$$(\sigma_1 - \sigma_3)\cos\theta\sin\theta = \tau \tag{3.79}$$

We consider once again a classical trigonometric identity:

$$\sin 2\theta = 2\cos\theta\sin\theta \tag{3.80}$$

Combining Eqs (3.79) and (3.80), we get

$$\tau = \frac{\sigma_1 - \sigma_3}{2}\sin 2\theta \tag{3.81}$$

Let us re-organise and square Eq. (3.78):

$$\left(\sigma_n - \frac{\sigma_1 + \sigma_3}{2}\right)^2 = \left(\frac{\sigma_1 - \sigma_3}{2}\cos 2\theta\right)^2 \tag{3.82}$$

We now square Eq. (3.81) and add it to Eq. (3.82) to derive the following equation:

$$\left(\sigma_n - \frac{\sigma_1 + \sigma_3}{2}\right)^2 + \tau^2 = \left(\frac{\sigma_1 - \sigma_3}{2}\right)^2 \tag{3.83}$$

Eq. (3.83) is the equation of a circle whose centre is $\left(\frac{\sigma_1 + \sigma_3}{2}; 0\right)$ and whose radius is $\frac{\sigma_1 - \sigma_3}{2}$ in the (σ_n, τ) space; it is the characteristic equation of the 2D Mohr circle (Fig. 3.9).

We see that each point of the circle represents the pole of a specific plane, noted **N** in Fig. 3.9A, the plane whose normal makes an angle θ with $\vec{\sigma}_1$ or, conversely and adopting a more usual convention in structural geology, the plane making an angle β with $\vec{\sigma}_1$ (Figs 3.8 and 3.9A). Note that angles in the physical space are doubled in the Mohr space. The coordinates of **N** in the Mohr space are the normal stress (Eq. (3.78)) and the shear stress (Eq. (3.81)) acting on this specific plane and given in abscissa and ordinate, respectively. The vector defined from the origin of the Mohr diagram to **N** is $\vec{\sigma}_n + \vec{\tau}$, hence the stress vector, $\vec{\sigma}$. We also note that compressive/tensile normal stresses and counter-clockwise/clockwise shear stresses are noted positive/negative according to Mohr diagram conventions (Fig. 3.9B). For the sake of clarity, it is preferable not to expand further the discussion but one might remark that the latter sign convention is not systematically consistent with the one adopted for the shear components of the stress tensor (cf. Twiss and Moores, 2001, p. 187). Finally, we remark that the abscissa value of the centre of the Mohr circle is equal to the mean stress (in 2D) and that its diameter corresponds to the differential stress (Eq. (3.83) and Fig. 3.9A).

A closer look at the Mohr circle reveals that shear stresses vanish for $2\theta = 0°$ and $2\theta = 180°$ (i.e. **N** plots on the abscissa; Fig. 3.9B), whereas the normal stress adopts its two extreme values, which are σ_1 and σ_3, respectively (Fig. 3.10). These two situations represent once again the particular cases the plane parallels one of the principal planes of stress. As the normal of the plane rotates away from $\vec{\sigma}_1$, the shear stress increases until reaching its maximum value for $2\theta = 90°$ ($\theta = 45°$). Visual inspection of the Mohr diagram shows that the allowed maximum shear stress value is equal to the radius of the Mohr circle, $\frac{\sigma_1 - \sigma_3}{2}$, or half the differential stress.

As 2θ increases further, the shear stress decreases until reaching 0 for $2\theta = 180°$ ($\theta = 90°$). The geometrical configurations represented by the lower half of the Mohr circle (i.e. for 2θ ranging between $180°$ and $360°$) are mirror images, according to the $\vec{\sigma}_1$ axis, of the ones represented by the upper half (Fig. 3.9B). This symmetry implies that the absolute value of the acute angle between $\vec{\sigma}_1$ and **N** is conserved between the two configurations of such symmetrical situations (i.e. angles being defined at $\pm 360°$, the case $2\theta = 300°$, symmetrical to the case $2\theta = 60°$, can also be defined by $2\theta = -60°$). Hence, only the sign of the shear stress, or conversely the sense of potential rotation imposed by the shear couple, differs between two symmetrical configurations. The Mohr circle is mostly used to

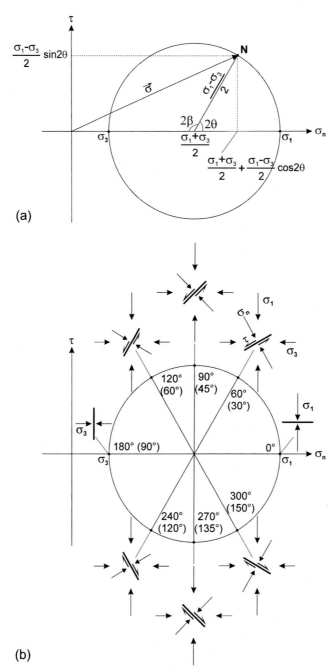

FIG. 3.9 (A) Meaning of Mohr circle and associated stress quantities (see details in text). In particular, **N** represents the pole of a given plane, whose coordinates in the Mohr space are respectively its corresponding normal and shear stress values. The angles θ and β are the angle between the normal of the plane and $\vec{\sigma}_1$ and the angle between the plane and $\vec{\sigma}_1$, respectively (see Fig. 3.8). (B) Representation of normal and shear components as a function of plane orientation, given as 2θ in the Mohr space (the values in parentheses correspond to θ in the physical space). Note that the lower half of the Mohr circle depicts situations symmetrical to the ones of the upper half.

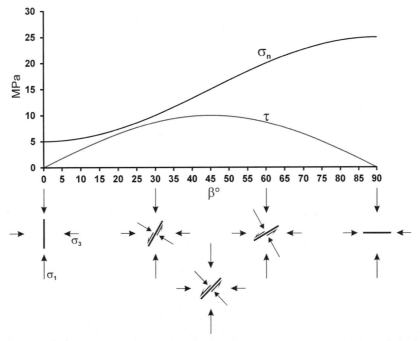

FIG. 3.10 Variation in normal and shear stress as a function of β, the acute angle separating the plane and $\vec{\sigma}_1$. The present graphs were generated based on Eqs (3.78) and (3.81) and assuming $\sigma_1 = 25$ MPa and $\sigma_3 = 5$ MPa.

determine absolute stress values and, in general, only its upper half is shown, as in the following text.

Deriving the equations of the 3D Mohr circle is a far more laborious exercise and does not add much to the comprehension of the physical meaning of Mohr circles. The details on the mathematical development of the 3D Mohr circle can be found at pp. 35–38 of Jaeger et al. (2007). We restrict here the discussion to the geometrical interpretation of the tool. The 3D Mohr circle can be viewed as a composition of three 2D Mohr circles, each circle representing one of the three principal planes of stress (Fig. 3.11). When the normal of the plane belongs to one of the principal planes of stress, its pole plots on the corresponding circle in the Mohr space. If the normal of the plane is not parallel to any of the three principal planes of stress, its pole falls somewhere in the area delimited by the three Mohr circles, i.e. grey area in Fig. 3.11B, white areas being excluded like in the case of the 2D Mohr circle.

In order to understand the construction of **N** in the Mohr diagram, let us focus first on the stress space (Fig. 3.11A). The head of the normal vector remains on the unit sphere regardless of the attitude of the plane. The orientation of the normal is given by three angles in their respective principal planes of stress; each of the three angles is found by rotating the normal along a cone whose axis of symmetry corresponds to one of the coordinate axis (or conversely to one of the principal stress directions). For example, rotation

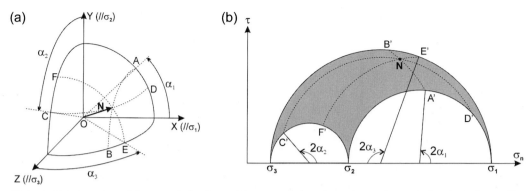

FIG. 3.11 (A) Orientation of the normal of a plane with respect to the principal axes of stress in the physical space (the principal axes of stress coincide with the axes of the coordinate system). (B) Counterparts in the Mohr space of the geometrical entities shown in (A). See text for detailed explanations on the construction of the pole **N** in the 3D Mohr circle.

about the X-axis (i.e. σ_1 axis) brings the normal in the $(\vec{\sigma}_1, \vec{\sigma}_2)$ plane where α_1 is measured. Similarly, α_2 and α_3 are determined after rotation about the Y-axis (i.e. σ_2 axis) and the Z-axis (i.e. σ_3 axis), respectively. In more detail, rotation about the X-axis brings the head of the normal vector to the point A in the $(\vec{\sigma}_1, \vec{\sigma}_2)$ plane such that OA and $\vec{\sigma}_1$ make an angle of α_1 (Fig. 3.11A). According to the principles described previously (e.g. Fig. 3.9A), the counterpart of A in the Mohr space, A', is to be found on the σ_1–σ_2 circle at an angle of $2\alpha_1$ from σ_1 (Fig. 3.11B). We rotate once again the normal vector about the X-axis but in the opposite direction until reaching the $(\vec{\sigma}_1, \vec{\sigma}_3)$ plane, now the head of the vector coincides with point B.

Let us examine the trajectory of the head of the normal vector from A to B. By definition, the motion from A to B follows a rotation about $\vec{\sigma}_1$; thus, the motion occurs along a plane parallel to the $(\vec{\sigma}_2, \vec{\sigma}_3)$ plane; that is, the path from A' to B' in the Mohr space corresponds to the arc of a circle concentric with the σ_2–σ_3 circle. It is relatively easy to find graphically and draw this circle containing A'. Its intersection with the σ_1–σ_3 circle determines B' (Fig. 3.11B). Following the same procedure, we can locate C' and E' and draw the arcs C'–D' and E'–F' in the Mohr space. The intersection between the three arcs determines the pole of the plane **N** and its corresponding stress values.

3.2 Fracture mechanics

3.2.1 Some words about linear elasticity

When loaded, rocks (and other materials) deform elastically before reaching failure. *Elasticity* is an ideal deformation mechanism, which implies full recovery of the initial shape of the deformed body after removal of the applied loads, in contrast to *plasticity*, where deformation is irrecoverable. A classical example of elastic behaviour is given by a spring recovering to its initial length after being stretched and then released. The latter example illustrates the main characteristic of an elastic body: its unloading path, as represented in

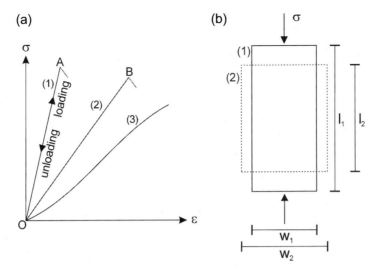

FIG. 3.12 (A) Elastic behaviour of three hypothetical materials as represented in a stress–strain plot. Materials (1) and (2) are linear elastic and material (3) is elastic but non-linear. Loading and unloading paths are shown only for material (1). Points *A* and *B* indicate elastic limits after which plastic yield begins, i.e. part of the deformation becomes irreversible. Note that material (1) presents a Young's modulus higher than the one of material (2), as suggested by the slopes of the loading/unloading paths. (B) Schematic representation of elastic deformation of a cylindrical rock sample during uniaxial loading in the laboratory. (1) and (2) indicate initial and deformed shapes of the sample, respectively (for the sake of clarity, deformation has been highly exaggerated).

a force/stress vs. displacement/strain plot, follows backwards the previous loading path (Fig. 3.12A). Thus, elastic deformation is characterised by energy or, more specifically, work conservation during loading/unloading cycles. During unloading, an elastic body returns the energy previously furnished to it in the course of the loading phase, whereas for a plastic body, energy is lost, consumed by (either brittle or ductile) mechanical yield of the material.

Linear elasticity, sometimes referred to as Hookean behaviour, is probably one of the most useful mechanical models. The concept was initially introduced by the famous English experimentalist, Robert Hooke (1635–1703) in 1676 as an enigmatic not to say a facetious anagram, which he clarified 3 years later, and owes its progressive development until the 19th century to various outstanding scientists (cf. Section 3.1.1). Linear elasticity represents the simplest model for elasticity by assuming linear loading/unloading paths (Fig. 3.12A) and the instantaneous response of the loaded/unloaded body. In other words, the deformation of a linearly elastic object is proportional to the load but independent of time. The latter statement is summarised in its simplest form by *Hooke's law*:

$$\sigma = E\varepsilon \tag{3.84}$$

where ε is strain and E is *Young's modulus*, the coefficient of proportionality between stress and strain.[a]

---

[a]The module of elasticity is named after Thomas Young (1773–1829), who advanced its first formal definition; however, the definition in use was given by Claude-Louis Navier (1785–1836).

Many elastic materials are not linear elastic and show complex loading/unloading paths (Fig. 3.12A) and/or time-dependent responses to loads (e.g. time lag between loading and deformation). The response of rocks to loading approximates Hookean behaviour, especially in laboratory experiments and for strains lower than 1% and, in general, at relatively low temperatures. Therefore, it appears that linear elasticity is appropriate to describe low strain deformation of the upper crust. Rocks typically exhibit Young's modulus values in the order of 10s of GPa. Thus, for strains of <1%, one expects stress magnitudes ranging from 10s to 100s of MPa, which correspond well to tectonic stress magnitudes.

In order to discuss further aspects of linear elasticity, let us first consider a typical experimental set-up, where a cylindrical rock sample is submitted to a uniaxial stress state (Fig. 3.12B), with all components of stress but the one parallel to the axis of the sample equal to zero, i.e. $\sigma_a = \sigma$ and $\sigma_r = 0$, where σ_a and σ_r are axial and radial stress components, respectively. Furthermore, we assume isotropic material. The axial symmetry of the system allows for simplification of the problem and derivation of the equations in 2D.

By definition, axial strain is given by

$$\varepsilon = \varepsilon_a = \frac{l_2 - l_1}{l_1} \tag{3.85}$$

Once again, we cannot avoid the eternal discussion on the vexing sign conventions. Eq. (3.85) shows that axial strain is negative, meaning that shortening of the sample is associated with negative strain (alternatively, positive strain results in stretching). By convention, Young's modulus is positive. Hence, according to Eq. (3.84), the compressive axial stress represented in Fig. 3.12B has to be negative! For the sake of consistency, we need to use here the 'engineering sign convention' and not the 'geologic sign convention' (cf. Section 3.1.2). The reader may understand now why some authors feel reluctant to use the 'geologic sign convention' (see discussion in e.g. Pollard and Fletcher, 2005, p. 210).

In its original expression, Hooke's law describes only the deformation of the sample in the direction of application of the load. However, experimental results show that elastic materials respond to axial compression by radial expansion as well (or conversely to axial extension by radial shortening). Considering isotropic material radial compression results evidently in axial expansion.

According to Fig. 3.12B, the radial strain is written as

$$\varepsilon = \varepsilon_r = \frac{w_2 - w_1}{w_1} \tag{3.86}$$

and it relates to the axial strain through

$$\nu = -\frac{\varepsilon_r}{\varepsilon_a} \tag{3.87}$$

where ν is *Poisson's ratio*, the second constant required to quantify elastic behaviour.

From Eq. (3.87), we see that the strain perpendicular to the direction of application of the load is proportional to the strain parallel to it. The negative sign indicates positive

Poisson's ratio values and ν varies from 0 (i.e. no Poisson's effect) to 0.5 for ideally *incompressible* materials. 'Incompressible' means that the volume of the sample remains constant during deformation and not specifically that the sample is rigid, i.e. does not respond to loading. Poisson's ratios of $\sim$0.2 to $\sim$0.4 are typical for rocks.

In the three-dimensional space, a more complex mathematical formulation than Hooke's law is required to describe elasticity. The *generalised Hooke's law* can be written as a set of linear equations:

$$\sigma_{ij} = \sum_{k=1}^{3} \sum_{l=1}^{3} c_{ijkl} \varepsilon_{kl} \tag{3.88}$$

where $i,j = 1,2,3$, and $i = j$ (and $k = l$) and $I \neq j$ (and $k \neq l$) indicate normal and shear components, respectively.

The c_{ijkl} coefficients are called *stiffnesses* and are linearly related to the elastic constants. The above set of linear equations can be written in matrix form (see details in e.g. Means, 1976), where the stiffnesses become the components of the *stiffness tensor*. Note that the stress and strain components are grouped in column vectors and not in array matrices. The set of Eq. (3.88) involves nine stress components and nine strain components. Therefore, the stiffness tensor is a 'monster' of 81 components in the general case and for anisotropic material. Fortunately, and considering the inherent symmetries of the system, the set of Eq (3.88) reduces for isotropic materials to the more pleasant form:

$$\sigma_{ij} = \frac{E}{1+\nu} \varepsilon_{ij} + \frac{E\nu}{(1+\nu)(1-2\nu)} \sum_{k=1}^{3} \varepsilon_{kk} \delta_{ij} \tag{3.89}$$

where $i,j = 1,2,3$ and δ_{ij} is the *Kronecker delta* with $\delta_{ij} = 0$, if $i \neq j$, and $\delta_{ij} = 1$, if $i = j$. Therefore, Eq. (3.89) represents a set of six equations, i.e. three equations furnishing the normal stress components and three other equations for the shear stress components, and nine independent stiffnesses.

In particular, if we write the stress tensor in the system of coordinates of its eigenvectors, the set of Eq. (3.89) simplifies to three equations giving the expressions of the three principal stresses (remember that shear components vanish; see Eq. (3.22)):

$$\sigma_1 = \frac{E}{(1+\nu)(1-2\nu)} [(1-\nu)\varepsilon_1 + \nu(\varepsilon_2 + \varepsilon_3)] \tag{3.90}$$

$$\sigma_2 = \frac{E}{(1+\nu)(1-2\nu)} [(1-\nu)\varepsilon_2 + \nu(\varepsilon_3 + \varepsilon_1)] \tag{3.91}$$

$$\sigma_3 = \frac{E}{(1+\nu)(1-2\nu)} [(1-\nu)\varepsilon_3 + \nu(\varepsilon_1 + \varepsilon_2)] \tag{3.92}$$

where repeated indexes are more conveniently replaced by single ones.

The two elastic constants, E and ν, are sufficient to describe the linear elastic behaviour of an isotropic material. However, it is common to use another equivalent pair of elastic constants, G and λ, called traditionally *Lamé's constants*.

The constant G is the *elastic shear modulus* and is written as a function of E and ν:

$$G = \frac{E}{2(1+\nu)} \tag{3.93}$$

As its name indicates, the shear modulus is used to relate shear stress to shear strain. Eq. (3.93) shows that $G = E/2$ for perfectly compressible material (i.e. $\nu = 0$) and $G = E/3$ for perfectly incompressible material (i.e. $\nu = 0.5$).

The second Lamé's constant, λ, is written as

$$\lambda = \frac{E\nu}{(1+\nu)(1-2\nu)} \tag{3.94}$$

Examination of Eq. (3.94) reveals that $\lambda = 0$ for perfectly compressible material (i.e. $\nu = 0$) and $\lambda \to \infty$ for perfectly incompressible material (i.e. $\nu \to 0.5$).

Combining Eqs (3.89), (3.93) and (3.94), we find the simplified expression of linear elasticity for isotropic solids:

$$\sigma_{ij} = 2G\varepsilon_{ij} + \lambda \sum_{k=1}^{3} \varepsilon_{kk}\delta_{ij} \tag{3.95}$$

and demonstrate one of the advantages of using Lamé's constants in the analytical treatment of linear elasticity theory.

We finally introduce an additional but important elastic parameter, the *bulk modulus*, K. The bulk modulus (or *incompressibility*) quantifies how much an elastic material resists a change in pressure (or isotropic stress). It relates the infinitesimal change in pressure, Δp, to infinitesimal volumetric strain through

$$K = \Delta p / \frac{\Delta V}{V} \tag{3.96}$$

where V is initial volume and ΔV is the change in volume. The inverse of K is the *compressibility*.

The bulk modulus can also be expressed as a function of E and ν (see details in e.g. Pollard and Fletcher, 2005, p. 297–298):

$$K = \frac{E}{3(1-2\nu)} \tag{3.97}$$

We remark that for a material approaching perfectly incompressible behaviour, i.e. $\nu \to 0.5$, K tends to an infinite value, in agreement with its physical meaning. For a perfectly compressive material (i.e. $\nu = 0$), $K = E/3$.

3.2.2 Frictional sliding

Before pursuing our journey through continuum mechanics, we make a short detour to explore a particularly relevant aspect to faulting: *frictional sliding*. Frictional sliding is a physical process involving forces resisting the relative motion of two bodies in contact, i.e. *frictional forces*. For example, we all have experience of sliding objects (e.g. a table) resting on a surface (e.g. the floor of our kitchen). We intuitively know that we need to overcome frictional forces to slide an object on a surface. We also know we need to push or pull with sufficient force for the object to start sliding and the heavier the object the more demanding is the effort.

The very first scientific study on friction was by Leonardo da Vinci (1452–1519), who concluded that resistance to sliding was proportional to weight but independent of contact area (Popova and Popov, 2015). Unfortunately, his findings remained hidden in his numerous notes for centuries. The French physicist, Guillaume Amontons (1663–1705) rediscovered and further developed da Vinci's early concept of friction (Amontons, 1699). The major contribution of Amontons to the physics of friction resides in the proportionality between frictional (or tangential) force and normal force, named later *Amontons' law*.

To understand the physical meaning of Amontons' law, let us consider the simple experiment depicted in Fig. 3.13, where only the force of gravity (i.e. $\vec{W}$) acts on a body resting on a plane, all other forces, like atmospheric pressure, being negligible. If the plane is horizontal, gravity acts perpendicular to the contact surface; therefore, there is no force component parallel to it. If we incline progressively the plane, the component of force tangent to the contact surface (i.e. $\vec{W}_x$) increases whereas the normal component (i.e. $\vec{W}_y$) decreases. The body remains at its initial position on the plane until the dip of the plane reaches a critical angle for which it starts to slide down the slope. This critical angle, $\phi = \phi_S$, is the *angle of static friction*.

We have shown from simple geometrical considerations that sliding initiates when ϕ reaches the critical value ϕ_S; this condition corresponds to a specific ratio between tangent and normal forces. Let us now search for a mathematical expression of the latter statement. According to Newton's Third Law, $\vec{W}_x$ and $\vec{W}_y$ are balanced by opposite forces. These are respectively the normal reaction of the plane, $\vec{F}_N$, and the frictional resistive force tangent to it, $\vec{F}_T$, such that

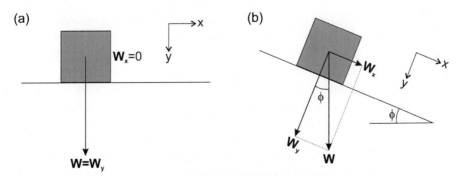

FIG. 3.13 Force components for a body resting on (A) a horizontal plane and (B) an inclined plane. Progressive increase in plane dip results in an increase in W_x but a decrease in W_y until the body starts sliding for a critical value of ϕ.

$$W_x = -F_T \tag{3.98}$$

$$W_y = -F_N \tag{3.99}$$

Considering the trigonometric relationships shown in Fig. 3.13B, the two components of the gravity force are.

$$W_x = W \sin\phi \tag{3.100}$$

$$W_y = W \cos\phi \tag{3.101}$$

Note that ϕ varies between $0°$ and $90°$.

Combining Eqs (3.98)–(3.101), we get

$$F_T = \tan\phi\, F_N \tag{3.102}$$

We previously concluded that sliding initiates when $\phi = \phi_S$, and thus, when $\tan\phi = \tan\phi_S$. Alternatively, the body remains at rest for $\phi < \phi_S$ or $\tan\phi < \tan\phi_S$. Using Eq. (3.102), these two conditions can be reformulated as a function of the normal and tangential forces:

$$\text{at rest (static)} \quad \frac{F_T}{F_N} < \tan\phi_S \tag{3.103}$$

$$\text{sliding initiates} \quad \frac{F_T}{F_N} = \tan\phi_S \tag{3.104}$$

Introducing the *coefficient of static friction*

$$\mu_S = \tan\phi_S \tag{3.105}$$

we write Amonton's law in its usual form:

$$\text{at rest (static)} \quad F_T < \mu_S F_N \tag{3.106}$$

$$\text{sliding initiates} \quad F_T = \mu_S F_N \tag{3.107}$$

To conclude with Amontons' law of static friction: the law tells us that *(1) the coefficient of friction equals the ratio between tangential and normal forces, (2) during (chiefly tangential) loading the coefficient of friction increases until reaching a critical value, the coefficient of static friction, for which sliding initiates.* Turning back to the illustrative example we devised in the introduction, the latter sentence is the translation in physical terms of our common life experience of shuffling furniture around the house.

Amonton's law represents a general and empirical theory of friction, which is still valid at the first order present-day. The detailed physics of friction, in particular at the micro scale, remain poorly known and are a vivid field of research, named tribology, highly relevant for engineering, material science and earthquake mechanics. To note, the coefficient of friction does not remain constant after sliding initiates but stabilises to a lower value, the *coefficient of dynamic (or kinetic) friction*, when the sliding velocity is constant, a result beyond reach for experimental facilities at the time of Amontons.

A modern and usual expression of Amontons' law as a function of stresses is as follows:

$$\text{static} \quad \tau < \mu_S \sigma_N \tag{3.108}$$

$$\text{sliding initiates} \quad \tau = \mu_S \sigma_N \tag{3.109}$$

Results from numerous experiments conducted on rocks in the laboratory point to μ_s values ranging commonly between ~ 0.5 and ~ 0.8 (Pollard and Fletcher, 2005). In addition, lab experiments show the linear relationship of Amontons to be incomplete. A more appropriate formulation for rocks (and granular materials in general) is.

$$\text{sliding initiates} \quad \tau = C_0 + \mu_S \sigma_N \tag{3.110}$$

where C_0 is the *cohesion* (or *frictional strength*) of the contact surface or discontinuity.

The latter parameter corresponds to the *shear strength* of the discontinuity, i.e. the level of shear stress for which sliding initiates, in absence of normal pressure across it. Typical cohesion values for rock surfaces range between few MPa at relatively low confining pressures to few tens of MPa otherwise.

Eq. (3.110) represents a line, often called the *friction line*, in the Mohr space (Fig. 3.14). By definition, the coordinates of any point below the friction line meet the relationship:

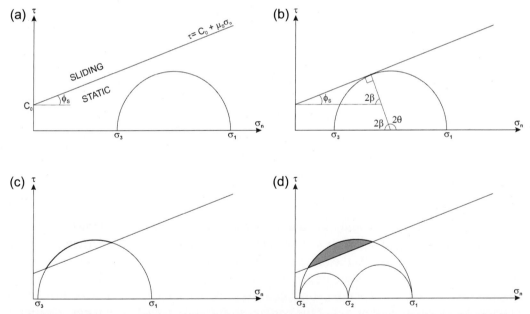

FIG. 3.14 Friction law as represented in Mohr diagrams. (A) The Mohr circle remains below the friction line; therefore, no planar discontinuity can be activated as sliding surface; (B) the Mohr circle is tangent to the friction line and sliding occurs only for optimally oriented discontinuities making an angle of β with $\vec{\sigma}_1$ (by symmetry sliding occurs also on discontinuities making an angle of $-\beta$); (C) part of the Mohr circle crosses the friction line and sliding occurs exclusively on all discontinuities whose directions are represented by the arc above the friction line; (D) similar to the case depicted in (C) but for a 3D stress state, sliding happens only for all discontinuities whose directions are represented by the grey cap above the friction line. All other symbols are detailed in the text.

$\tau < C_0 + \mu_S \sigma_N$. Thus, the condition for sliding is never reached for any planar discontinuity submitted to a stress state whose Mohr circle plots below the friction line (Fig. 3.14A). If the friction line is tangent to the Mohr circle, only two optimally oriented planes making angles of $\pm\beta$ with $\vec{\sigma}_1$ slide (Fig. 3.14B). Examining the right triangle depicted in Fig. 3.14B, we relate β to the angle of friction, ϕ_S:

$$2\beta + \phi_S + 90° = 180° \tag{3.111}$$

$$\text{or} \quad \beta = 45° - \frac{\phi_S}{2} \tag{3.112}$$

Eq. (3.112) demonstrates that sliding on a discontinuity at $\beta = 45°$, i.e. when the shear stress reaches its maximum value (see Fig. 3.10), is only possible for frictionless contacts or, conversely, when the friction line parallels the abscissa. In the general case, β is less than 45°. Taking $\phi_S = 30°$ as a reasonable estimate for most granular materials, the angle between the optimally oriented plane and $\vec{\sigma}_1$ is (coincidently) equal to 30°.

When the Mohr circle crosses the friction line (Fig. 3.14C), sliding occurs on less optimally oriented planes, represented by the arc above the friction line in 2D (Fig. 3.14C). In 3D, the collection of sliding planes is naturally represented by the circular cap above the friction line (Fig. 3.14D), implicitly meaning that the friction condition depends on the three principal axes of stress.

Based on laboratory experiments, various empirical friction laws similar to Eq. (3.110) have been proposed. *Byerlee's Law* is certainly the most utilised friction law in geology. The American geophysicist, James Byerlee (1927–2015) compiled a large number of measurements of static friction, including his original results (Byerlee, 1978) carried out on a wide range of lithologies at different stress conditions. Practically, the experiments were conducted on pre-cut rock samples submitted to variable confining pressures (thus variable normal stress) and μ_S was determined as the ratio between the shear stress required to initiate sliding and the confining stress. Byerlee concluded on two friction behaviours depending on σ_N/depth:

$$\tau = 0.85\,\sigma_N \quad 0 \le \sigma_N \le 200\,\text{MPa} \tag{3.113}$$

$$\tau = 50 + 0.6\,\sigma_N \quad 200 \le \sigma_N \le {\sim}\,1700\,\text{MPa} \tag{3.114}$$

where stresses are in MPa, and as usual positive values indicate compressive stresses.

Eq. (3.113) applies to relatively shallow depths down to $\sim$8 km, where friction appears to be particularly sensitive to the roughness of the contact surface and not only to confining pressure. Eq. (3.114) is much better constrained by the data used by Byerlee, suggesting that friction is solely dependent on confining pressure at relatively great depths, where efficient locking of the asperities of the contact surface takes place. The latter equation is presumably valid for the entire brittle crust at relatively great depths and assumed to be valid for the brittle part of the mantle as well.

3.2.3 Brittle failure

After having introduced the notion of friction, we now proceed with the deformation of continuous media and, in particular, explore the mechanisms of brittle deformation when the elastic limit of the material is exceeded (Fig. 3.12A). Based on results from laboratory experiments, the two basic equations describing the onset of fracture are

$$\sigma_3 = -T \tag{3.115}$$

where $T \geq 0$ is *uniaxial tensile strength* and

$$\sigma_1 = \sigma_c + k\sigma_3 \tag{3.116}$$

where $\sigma_c \geq 0$ is *uniaxial compressive strength* (i.e. when $\sigma_3 = 0$) and $k \geq 0$ is a constant.

The first of the two equations predicts tensile fracture of the material when the minimum principal stress is negative and reaches the critical value $-T$, which is a characteristic of the material under scope. The fracture propagates perpendicular to $\vec{\sigma}_3$ and opens as a mode 1 fracture (Fig. 3.15A and Fig. 2.1). Eq. (3.116) addresses the more general case of biaxial loading and involves fracture under compression, potentially resulting in the onset of shear fractures (Fig. 3.15B). The criterion proposes a linear relationship between σ_1 and σ_3, which is valid at the first-order, though rocks exhibit very often non-linear relationships. The two equations neglect the potential influence of the intermediate principal stress σ_2 in the fracture process. In many cases, σ_2 also plays a role in the brittle deformation of rocks; however, the latter is modest and can be ignored when dealing with natural fractures.

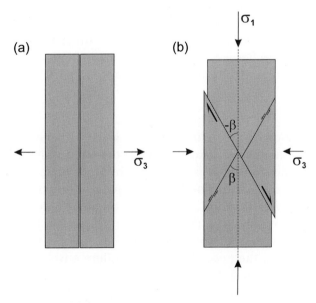

FIG. 3.15 Fracture in (A) tension and (B) shear.

In 1773, the French military engineer and physicist, Charles-Augustin Coulomb (1736–1806) presented a note on the strength of construction materials and stability of structures (Coulomb, 1776). This extensive note summarised various experimental and theoretical results and, in particular, his research on brittle failure by shear. Making an analogy with Amontons' law, Coulomb concluded that the shear force required to fracture a solid was opposed by (1) a friction-like force parallel to the potential plane of fracture and (2) the shear strength of the material. A modern expression of *Coulomb fracture criterion* (sometimes called *Navier–Coulomb fracture criterion*) is

$$\tau = C_i + \mu_i \sigma_N \tag{3.117}$$

where $\mu_i = \tan\phi_i$ is the *coefficient of internal friction*, ϕ_i is the *angle of internal friction*, and C_i is customarily termed *cohesion* (or *inherent shear strength*). In order to avoid potential confusion with cohesion of contact surfaces and for the sake of consistency, we will call informally the latter parameter 'internal cohesion'. The strength of materials is insensitive to the sense of shear; thus, Eq. (3.117) is generally employed with $\tau \geq 0$. Laboratory tests point to typical values of $\sim$10 to 50 MPa and $\sim$30° for internal cohesion and angle of friction of rocks, respectively.

Despite the mathematical similarities between Amontons' law and Coulomb fracture criterion, these two empirical laws describe distinct physical phenomena and conditions. In Amontons' law, the contact surface, which we assimilate to a fracture plane, is a physical object responding with its own mechanical properties (i.e. cohesion and coefficient of static friction). In Coulomb fracture criterion, stresses act on virtual planes within a continuous solid, and the latter responds according to its mechanical properties (i.e. 'internal cohesion' and internal friction). Coulomb fracture criterion does not furnish any information on the location of the potential shear fracture nor on the exact mechanisms leading to rupture. It only predicts the stress condition for which the material fails and the orientations of the shear fractures (i.e. $\pm\beta$; Fig. 3.15B). Similar to frictional sliding (Eq. 3.112), the angle between the fracture plane and $\vec{\sigma}_1$ is given by

$$\beta = 45° - \frac{\phi_i}{2} \tag{3.118}$$

Eq. (3.117) corresponds to the *failure envelope*, which is, in the present case, a line with slope ϕ_i and intersecting the ordinate at $\tau = C_i$ in the Mohr space (Fig. 3.16A). Failure is predicted when the failure envelope is tangent to the Mohr circle. In other words, the pole of the newly formed shear fracture (and of the corresponding stress vector; see Fig. 3.9A) always belongs to the $(\vec{\sigma}_1, \vec{\sigma}_3)$ plane. That is, $\vec{\sigma}_2$ is parallel to the fracture plane and perpendicular to the shear stress and does not intervene in the failure process. Fracture nucleation requires energy consumption, which in turn implies stress drop (i.e. the Mohr circle shrinks). Therefore, the portion above the failure envelope is meaningless as no Mohr circle can cross the envelope.

Although the Coulomb fracture criterion was established for compressive stresses, it is theoretically possible to extend its corresponding failure envelope to the left of the

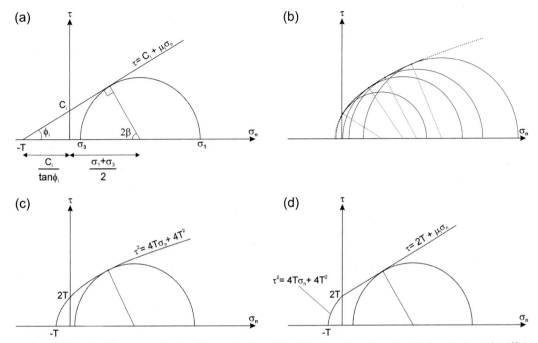

FIG. 3.16 (A) Coulomb fracture criterion; (B) construction of the failure envelope based on Mohr criterion; (C) Griffith fracture criterion; (D) modified Griffith fracture criterion. See text for explanation of symbols.

ordinate (Fig. 3.16A). Its intersection with the abscissa automatically gives the signed value of the tensile strength, *T*.

Let us now examine the largest triangle depicted in Fig. 3.16A. Considering usual trigonometric relationships, we write

$$\frac{\sigma_1 - \sigma_3}{2} = \left(\frac{\sigma_1 + \sigma_3}{2} + \frac{C_i}{\tan\varphi_i} \right) \sin\varphi_i \tag{3.119}$$

and after re-arranging the above equation, we get

$$\sigma_1 = \frac{2C_i \cos\varphi_i}{1 - \sin\varphi_i} + \frac{1 + \sin\varphi_i}{1 - \sin\varphi_i}\sigma_3 \tag{3.120}$$

Eq. (3.120) is similar to (3.116), where

$$k = \frac{1 + \sin\varphi_i}{1 - \sin\varphi_i} \tag{3.121}$$

$$\sigma_c = 2C_i\sqrt{k} \tag{3.122}$$

Therefore, it is theoretically possible to derive uniaxial compressive strength from internal cohesion and internal friction.

We now examine the smallest triangle in Fig. 3.16A and get

$$T = \frac{C_i}{\tan\varphi_i} \tag{3.123}$$

As already mentioned, average angles of internal friction are less than 45° for rocks and also for granular materials in general. Thus, Eq. (3.123) predicts $T > C_i$ in clear contradiction with experimental results, where the tensile strength was found to be approximatively equal to half of the internal cohesion. In qualitative terms, the linear failure envelope appears to project tensile strength values too far to the left of the ordinate (Fig. 3.16A).

In order to correct for this serious shortcoming, several solutions were proposed after Coulomb. Mohr (1900) assumed a more general expression for the failure envelope and proposed to determine it experimentally as the curve tangent to all Mohr circles associated with failure of the tested material (i.e. *Mohr* or *Mohr-Coulomb criterion*; Fig. 3.16B). The constructed failure envelopes confirmed the general validity of Coulomb criterion for compressive normal stresses and relatively low confining pressures, and extrapolation of the curved Mohr envelopes to the left of the ordinate led to more satisfactory tensile strength values.

With the advent of modern physics in the early 20th century, the English engineer Alan Arnold Griffith (1893–1963) investigated brittle failure from a theoretical point of view. Macroscopic rupture of materials had to be the natural consequence of breaking atomic bonds at the microscopic scale. However, atomic bounding theory predicted tensile strength values for solids in the order of the 10th of their respective Young's modulus or, in other words, three orders of magnitude higher than what was experimentally measured. To solve this paradox, Griffith (1921) postulated the existence of microscopic flaws in solids. He demonstrated, in particular, that stresses were sufficiently enhanced at the tips of optimally oriented elliptical cracks (i.e. *Griffith's cracks*) to overcome the strength of atomic bonds in the close vicinity of the tips, and to allow the cracks to extend and link until complete failure of the material (Griffith, 1924).

His analysis (see Jaeger et al., 2007, pp. 314–317 for detailed mathematical treatment) resulted in the *Griffith fracture criterion*, traditionally written in the space of the principal stresses:

$$\sigma_3 = -T \quad \text{if } \sigma_1 \leq -3\sigma_3 \tag{3.124}$$

$$(\sigma_1 - \sigma_3)^2 - 8T(\sigma_1 + \sigma_3) = 0 \quad \text{if } \sigma_1 > -3\sigma_3 \tag{3.125}$$

As for Coulomb and Mohr criteria, σ_2 does not play any role in the fracture process. The uniaxial compressive strength is found by examining the particular case of uniaxial compression, i.e. by setting $\sigma_3 = 0$ in Eq. (3.125):

$$\sigma_c = 8T \tag{3.126}$$

and predicted uniaxial compressive strength values remain merely below experimental values but of the correct order.

In the Mohr space, the failure envelope corresponding to Griffith criterion is a parabola (Fig. 3.16C):

$$\tau^2 = 4T\sigma_n + 4T^2 \tag{3.127}$$

Setting $\sigma_N = 0$ in Eq. (3.127), a particularly interesting outcome of Griffith fracture criterion is

$$C_i = 2T \tag{3.128}$$

in good agreement with experimental results.

Griffith criterion furnishes theoretical grounds to the previous works of Coulomb and Mohr and has the advantage of dealing with both shear and tensile fracture. For $\sigma_N \geq 0$, the criterion predicts shear fracture and its failure envelope becomes quasi-linear, mimicking the Coulomb criterion (Fig. 3.17). For $\sigma_N = \sigma_3 = -T$, the criterion resembles the empirical Eq. (3.115), and purely tensile, mode I, fracture is predicted. For $-T < \sigma_N < 0$, the criterion predicts *hybrid fracture*, in which rupture occurs in response to both shear and tension (Fig. 3.17). Experimental validation of this specific part of the failure envelope is still in progress (see discussion in Engelder, 1999), albeit significant advances have been made by Ramsey and Chester (2004). On one hand, experimentalists face arduous technical problems to reproduce the required stress conditions. On the other hand, it remains unclear whether hybrid fractures observed in the field (see Chapter 2, Section 2.2) nucleated at such or opened as mixed-mode fractures later after rupture.

McClintock and Walsh (1962) considerably improved Griffith's theory in postulating the elliptical cracks would close at relatively high confining pressures to act as frictional contacts within rock masses. Their analytical development led to the *modified Griffith fracture criterion*:

$$\left(\sqrt{1+\mu^2} - \mu\right)\sigma_1 - \left(\sqrt{1+\mu^2} + \mu\right)\sigma_3 = 4T\sqrt{1 + \frac{\sigma_{cl}}{T}} - 2\mu\sigma_{cl} \tag{3.129}$$

where μ is the crack wall friction and σ_{cl} is the minimum normal stress value required to close the cracks.

As confining pressure increases, Griffith's cracks close and σ_{cl} becomes negligible in Eq. (3.127), which in turn can be written as

$$\left(\sqrt{1+\mu^2} - \mu\right)\sigma_1 - \left(\sqrt{1+\mu^2} + \mu\right)\sigma_3 = 4T \tag{3.130}$$

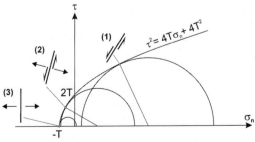

FIG. 3.17 The three types of fracture predicted by Griffith fracture criterion: (1) shear, (2) hybrid and (3) tensile fracture.

Eq. (3.130) corresponds in the Mohr space to

$$\tau = 2T + \mu\sigma_n \tag{3.131}$$

which is equivalent to Coulomb fracture criterion (Eq. (3.117)), with $C_i = 2\,T$ and $\mu_i = \mu$.

In brief, the modified Griffith fracture criterion can be viewed as a combination of Griffith criterion for negative normal stresses, with Coulomb criterion for positive normal stresses (Fig. 3.16D).

More recently, Hoek and Brown (1980) proposed an alternative empirical fracture criterion:

$$\sigma_1 = \sigma_3 + \sqrt{m\sigma_c\sigma_3 + s\sigma_c^2} \tag{3.132}$$

where m and s are dimensionless fitting parameters related to rock mass quality and mechanical properties, respectively.

The fit to experimental data of *Hoek and Brown criterion* (and of its updated versions) is satisfying, and the criterion is widely used in geosciences and more specifically in engineering geology, its own field of development.

3.2.4 Role of pore pressure

Until now, we have focused our discussion on fundamental aspects of frictional sliding and strength of solids and rocks following the main outcomes of laboratory work and mathematical analyses. One essential parameter pertaining to the in situ conditions of rocks underground is the pressure of the fluids they host in their pore space, namely *pore pressure*. The geotechnical engineer Karl von Terzaghi (1883–1963), born in Prague under the Austro–Hungarian Empire, was the first researcher to understand the relevance of pore pressure for soil stability (von Terzaghi, 1936). His findings were extended later to rock mechanics and tectonic problems (Hubbert and Rubey, 1959; Secor, 1965).

Pore pressure acts inside rock cavities and exerts outward-directed normal stresses, as opposed to the generally compressive stresses, which result from the natural loading of rocks (e.g. pressure of the rock column or tectonic loading). Assuming random distribution of spherical pores within the rocks, pore pressure leads to a decrease in stress magnitudes in all directions according to

$$\sigma'_i = \sigma_i - P \quad \text{with } i = 1,2 \text{ or } 3 \tag{3.133}$$

where P is pore pressure and σ'_i is the *effective stress*.

In short, the transformation simply consists of subtracting pore pressure to the stress the rock would experience in the absence of fluids, using the classical principle of superposition. A pleasant implication of it is that all the equations devised in this chapter remain valid to treat cases involving pore pressures, provided normal stress components are carefully changed to their effective counterparts.

In the Mohr space, the addition of pore pressure is equivalent to shifting the Mohr circle to the left at a distance P from its initial position (Fig. 3.18). The radius of the Mohr

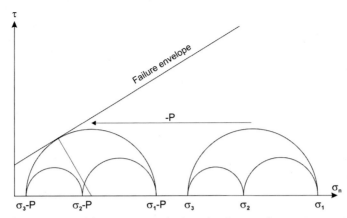

FIG. 3.18 Influence of pore pressure (*P*) on stress magnitude and rock strength.

circle, as well as the other relative distances between the principal stresses, remain untouched as pore pressure is by nature an isotropic stress state, whose addition or subtraction does not affect stress differences. If *P* is sufficiently elevated, the Mohr circle may touch the failure envelope, i.e. *pore pressure acts as a weakening factor in the brittle field of deformation.*

In the case of shear fractures, either nucleated according to e.g. Coulomb criterion or reactivated as frictional contacts according to Amontons law, pore pressure leads to decreasing normal stresses that work against shear, thus facilitating the occurrence of shear failure and displacement. In the case of tensile rupture, pore pressure reduces the effective minimum principal stress until it potentially reaches the (signed) value of the tensile strength of the rock. For a pre-existing discontinuity, pore pressure works against the normal stress that maintains it closed, and the discontinuity opens as a mode 1 fracture when pore pressure equals or exceeds the normal stress.

Further reading

Jaeger, J.C., Cook, N.G.W., Zimmerman, R., 2007. Fundamentals of Rock Mechanics. Wiley.

Means, W.D., 1976. Stress and Strain: Basic Concepts of Continuum Mechanics for Geologists. Springer-Verlag, New York.

Pollard, P.D., Fletcher, R.C., 2005. Fundamentals of Structural Geology. Cambridge University Press.

Ramsey, J., Chester, F., 2004. Hybrid fracture and the transition from extension fracture to shear fracture. Nature 428, 63–66.

Ramsay, J.G., Lisle, R.J., 2000. The Techniques of Modern Structural Geology, Volume 3: Applications of Continuum Mechanics in Structural Geology. Academic Press, London.

Timoshenko, S., 1953. History of Strength of Materials. McGraw-Hill, New York.

Twiss, R.J., Moores, E.M., 2001. Structural Geology, 7th printing of the 1992 ed. W. H. Freeman, New York.

4

Fault slip inversion methods

4.1 Background of fault slip inversion methods

4.1.1 Anderson's theory of faulting

Early in the 20th century, the Scottish geologist Ernest Masson Anderson (1877–1960) made a significant step towards the quantification of stress from observation and measurement of tectonic fractures (Anderson, 1905, 1951). Anderson and previous geologists remarked that fault dip was often correlated with type of faulting, the planes of normal, strike-slip ('wrench' in Anderson, 1951) and thrust faults being in general steeply inclined, nearly vertical and shallow dipping, respectively (Fig. 4.1). In addition, it was commonly observed that faults used to be arranged in pairs of analogous faulting type, the two faults being at nearly 60° from each other (i.e. *conjugate faults*). Based on the latter observation, Anderson postulated that natural faults were similar to the shear fractures predicted by Coulomb fracture criterion[a] (see Chapter 3, Section 3.2.3 and Fig. 3.15) and reproduced experimentally by Daubrée (1879) among others.

Thus, according to Anderson the symmetry of a pair of conjugate faults should reflect the orientations of the elements of symmetry of the stress responsible for rupture[b] or, in other words, the orientations of the principal axes of stress. Noteworthy, Gzovskii (1954), in Russia, and Hirayama and Kakimi (1965), in Japan, reached similar conclusions independently. The fact that pairs of conjugate faults exhibit very often attitudes consistent with a vertical axis of symmetry (Fig. 4.1) adds a strong constraint: stress configurations with one vertical principal axis of stress are the most usual ones in nature.

The two previous sentences constitute the background of *Anderson's theory of faulting*, which can be enunciated:

(1). *one principal axis of stress is always (nearly) vertical and the two other (nearly) horizontal in the brittle crust;*
(2). *faults form conjugate pairs with acute angles of ~60°, for each pair of conjugate faults $\vec{\sigma}_1$ bisects the acute angle, $\vec{\sigma}_3$ bisects the obtuse angle and $\vec{\sigma}_2$ parallels the intersection line between the two fault planes;*

[a]Anderson was aware of Coulomb's contribution but seemed not to have had access to his seminal paper published in 1776, and erroneously attributed the paternity of the criterion to Navier (1833), who was also known for his influential work on shear fracture of materials (see p. 3 in Anderson, 1951).

[b]Anderson applied (intuitively?) the main principle of symmetry formalised by the famous physicist and crystallography pioneer Pierre Curie (1859–1906): 'When given causes produce given effects, the elements of symmetry of the causes have to be found in the produced effects'. Translated after the biography of Pierre Curie written by Marie Skłodowska Curie (Curie, 1924).

Paleostress Inversion Techniques. https://doi.org/10.1016/B978-0-12-811910-5.00003-8

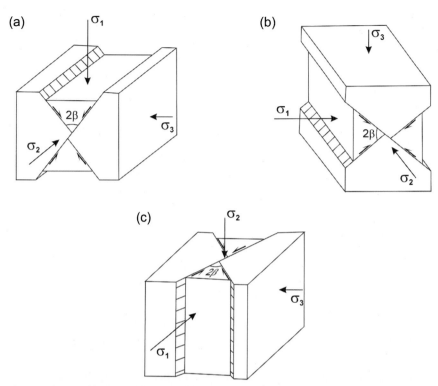

FIG. 4.1 Anderson's theory of faulting. (A) normal, (B) reverse and (C) strike-slip faulting and their respective principal stress orientations (for clearness the depicted configurations were intentionally drawn in violation of the rule of area/volume conservation between initial and deformed states).

(3). *as corollary of (1) and (2), reverse faults dip~30°, normal faults~60° and strike-slip faults are nearly vertical.*

The latter statement defines *Andersonian faults.* Fig. 4.1 shows that the only difference between the three types of Andersonian faults resides in orientation with respect to vertical. The characteristic symmetry of conjugate pairs of faults about the stress axes remains whatever faulting type is considered. For example, the configurations for reverse (Fig. 4.1B) and strike-slip (Fig. 4.1C) faulting are obtained after rotating 90° the configuration for normal faulting (Fig. 4.1A) about $\vec{\sigma}_2$ and $\vec{\sigma}_3$, respectively.

A stress state with vertical $\vec{\sigma}_1$, promoting normal faulting, is by extension categorised as *normal stress regime* (Fig. 4.1). In a similar manner, *reverse stress regime* and *strike-slip stress regime* indicate stress states with vertical $\vec{\sigma}_3$ and $\vec{\sigma}_2$, respectively. The reader should be aware that the word 'stress' is often overlooked when using these definitions in the literature, leading to potential ambiguity between stress regime and faulting regime.

A stress state with one vertical principal axis of stress is referred to as *Andersonian stress state.* Anderson advanced two physical arguments to support the prevalence of Andersonian stress states in the brittle crust. First, the Earth's surface represents a mechanical free surface (i.e. air or water is not mechanically coupled to the underlying rocks). By definition, shear stresses vanish on a free surface and one of the axes of principal stress is therefore orthogonal to it. Hence, at relatively shallow depths in the underground, one principal axis of stress is in general (nearly) vertical, except in areas of rough relief. Additionally, the weight of the rock column exerts a downwards pressure (i.e. *lithostatic pressure*), which naturally corresponds to one of the principal stresses. This second argument ensures that Andersonian stress states apply also to levels of the brittle crust well below the Earth's surface.

Anderson's theory of faulting provides a means to infer the orientations of the principal axes of stress according to measured fault attitudes and is solely based on symmetry considerations (Fig. 4.2). It is restricted to shear rupture surfaces, called *neoformed faults*, characterised by either pure strike-slip, for strike-slip faulting, or pure dip slip for normal or reverse faulting. The theory breaks down when it comes to predicting stress orientations for faults reactivated in stress conditions differing from the one that initiated them.

In general, no other stress parameter than the orientations of the principal stresses can be resolved. Exceptions to this rule are collections of coeval either normal or reverse dip slip faults with uniformly distributed strikes or, conversely, presenting a radial symmetry

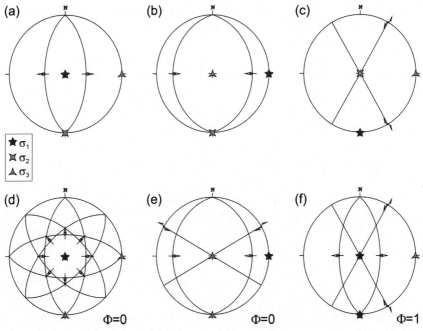

FIG. 4.2 Synthetic examples of the application of Anderson's theory of faulting. (A), (B) and (C) depict common cases of normal, reverse and strike-slip faulting, respectively, where only stress orientations can be resolved. For the more seldom cases of (D) radial extension, (E) transpressive stress regime and (F) transtensive stress regime, the value of the shape ratio, Φ, can also be estimated. Note that for the sake of simplification only few faults are depicted; in practice, robust determinations require statistically significant numbers of measurements. Projection is on the lower hemisphere of a Schmidt net.

according to the vertical axis, a geometrical configuration that automatically points to stress isotropy along one of the principal planes of stress. For example, a set of coeval normal dip slip faults, whose strikes are uniformly distributed, suggest isotropy along the horizontal $(\vec{\sigma}_2, \vec{\sigma}_3)$ plane or $\sigma_2 = \sigma_3$, that is a shape ratio (see Eq. (3.38) and Fig. 3.6) $\Phi = 0$. The latter stress state is often referred to as *radial extension*. Radial extension appears geologically sound to describe gravitational collapse of e.g. volcanic edifices or mountain chains. However, *radial compression*, involving uniformly distributed reverse faults, hence isotropy along the horizontal $(\vec{\sigma}_1, \vec{\sigma}_2)$ plane (i.e. $\Phi = 1$), seems rather unlike in nature.

Other particular situations where Anderson's theory may furnish additional information on the stress state are the ones for which coeval faults of distinct faulting type indicate similar directions for the stress axes. For instance, the feasibility of coeval reverse and strike-slip ruptures requires isotropy along the vertical $(\vec{\sigma}_2, \vec{\sigma}_3)$ plane, that is $\Phi = 0$ and σ_1 horizontal (Fig. 4.2E). Such a specific stress state is usually called *transpressive stress regime* or *transpression*[c]. Alternatively, nucleation of coeval normal and strike-slip faults occurs under *transtensive stress regime* or *transtension*, with $\Phi = 1$ and σ_3 horizontal (Fig. 4.2F).

Note that Anderson apparently did not envisage these particular cases, for which stress anisotropy could also be inferred from geometries of neoformed faults. They represent nevertheless direct implications of his theory, and are tentatively grouped here under the term 'Extended Anderson's theory of faulting'.

However, not all faults are Andersonian limiting to some extent the use of Anderson's faulting theory. For example, (nearly) horizontal faults are found in the field as e.g. *détachements* or as flats in ramp-and-flat systems (see definitions in Twiss and Moores, 2001). Conversely, (nearly) vertical faults with significant vertical components of slip are occasionally observed. Local accommodation of strain, anomalous pore pressures or local alteration of the stress field are the most invoked causes for these occurrences, though the exact mechanisms of large *détachement* normal faults are still debated (e.g. Collettini, 2011).

Polymodal faults, involving in general four coeval fault sets of neoformed conjugate faults arranged in orthorhombic geometry (Aydin and Reches, 1982; Krantz, 1988) represent another limitation to Anderson's theory. Together with laboratory tests under triaxial conditions (Reches and Dieterich, 1983), these observations challenge the validity of fracture criteria neglecting the influence of σ_2, hence of Coulomb criterion as well, and by extension that of Anderson's theory of faulting. One reasonably conceives triaxial stress states to be prevalent in nature and should expect plane strain deformation, as predicted by Anderson for neoformed faults, to be exceptional. Surprisingly, the number of reported

---

[c]The terms 'transpression' and 'transtension' refer traditionally to specific kinematics of strike-slip faults (see e.g. p. 177 in van der Pluijm and Marshak, 2004). In paleostress literature the two words are unfortunately used to characterise particular stress states. The author recommends the use of 'transpressive stress regime' and 'transtensive stress regime' in order to avoid potential confusion.

FIG. 4.3 An example of application and validation of Anderson's theory of faulting. The picture shows conjugate normal faults, reminiscent of Andersonian faults, in Permo-Triassic sediments (Xivares Beach, Asturias, Spain). The rock face is nearly perpendicular to fault strike and black arrows indicate a ∼40–50 cm thick sandstone bed offset by the faults. The fault to the right exhibits an apparent dip higher than that of the fault to the left, and the imaginary axis bisecting the acute angle between the two conjugate faults is moderately tilted with respect to vertical but nearly orthogonal to bedding. Rotation of the bedding to horizontal brings back the two normal faults to similar dips of ∼55° and $\vec{\sigma}_1$ to vertical, in agreement with Anderson's theory. Hence, the depicted configuration suggests normal faulting before large-scale folding of the formation, fact which is confirmed by additional observations at the same locality and well documented at the regional scale (the faulting and the folding are presumably Late Jurassic and Early Cenozoic, respectively, e.g. Lepvrier and Martínez-García, 1990, Granado et al., 2018).

convincing cases for polymodal faulting is to date merely low (see references in Healy et al., 2015), whereas Andersonian bimodal faults are continuously documented in the literature for more than one century. The reason for this apparent paradox is a highly relevant issue in fracture mechanics but its origin remains enigmatic. Empirical observations confirm nevertheless the general validity of Anderson's theory of faulting (see example in Fig. 4.3) and, keeping in mind the above limitations, its usefulness in tectonic studies.

4.1.2 Wallace–Bott hypotheses

a) Background

Anderson demonstrated how the geometrical arrangement of neoformed faults could be used to infer stress orientations. What about the slickenlines commonly found on fault surfaces (see Chapter 2, Section 2.3.2) and identified as 'tracks' of shear displacement by Anderson (1948)? Did they also relate to the stresses at the origin of the observed fault slips? The analogy between faulting and the experimental setups of Amontons and his successors was certainly in scientists' minds in the first half of the 20th century, and assuming parallelism between fault slip and shear stress was seemingly a very natural hypothesis.

The American earthquake geologist Robert E. Wallace (1917–2007) and, independently, the British geophysicist Martin H.P. Bott (1926–2018) made implicitly this latter hypothesis in two different studies. Wallace (1951) explored mathematically the variation of vectors of maximum shear stress on imaginary planes in function of stress orientation and anisotropy (see Chapter 3, Eq. (3.70)). Bott (1959) used a similar mathematical approach to investigate the feasibility of oblique slip under Andersonian stress states. Both authors concluded on the influence of stress anisotropy on shear stress, thus fault slip orientation (Fig. 4.4), a result that was far to be intuitive for the geological community at the time. The scheme proposed by Wallace and Bott was applicable to all kinds of preexisting discontinuities reactivated as faults and, therefore, differed markedly from Anderson's theory, which was bound to the study of neoformed faults presenting striae systematically perpendicular to the line of intersection of the two conjugate faults.

As pointed out previously, neither Wallace (1951) nor Bott (1959) stated explicitly the fundamental hypothesis of parallelism between shear displacement or, in geological terms, slickenline and shear stress. They apparently considered this hypothesis so obvious that they found redundant to enounce it clearly. *Wallace–Bott hypotheses*, i.e. the

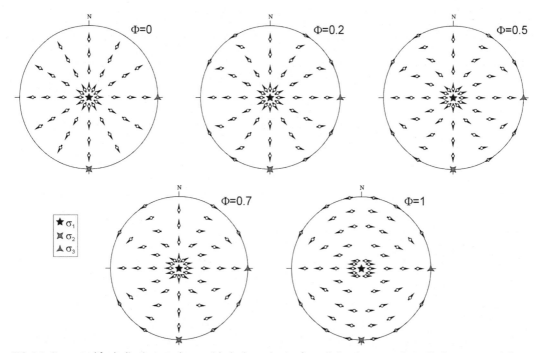

FIG. 4.4 Computed fault slips in accordance with the hypothesis of parallelism between shear displacement and shear stress of Wallace (1951) and Bott (1959). The results are for a normal stress regime with σ_3 striking *E*-W and five distinct values of the shape ratio, Φ. The case $\Phi = 0$ represents radial extension, where only dip slip is permitted. Note that the relative proportion of fault slips with marked strike-slip components increases with Φ. The results are presented as Hoeppner plots (Hoeppner, 1955), open dots correspond to the poles of the fault planes and arrows point to slip directions. Projection is on the upper hemisphere of a Schmidt net.

fundamental one or *main Wallace–Bott hypothesis* and a set of additional ones, were later formalised during the advent of fault slip inversion methods (Arthaud, 1969; Carey and Brunier, 1974; Carey, 1976) and are summarised as:

(1) *fault slip occurs parallel to the maximum shear stress;*
(2) fault surfaces are planar;
(3) deformation of the fault blocks is negligible;
(4) fault slip is small in comparison to the scale of observation;
(5) there are no stress perturbations and, in particular, no fault interactions;
(6) no rotation takes place along the faults during faulting;
(7) a uniform stress state prevails during one single tectonic event.

These hypotheses constitute the background of the methods of paleostress inversion of fault slip data. In particular, the main Wallace–Bott hypothesis is essential because it allows for relating faults to the mathematical concept of stress in a more complete way than Anderson's theory. Wallace–Bott hypotheses will be further discussed in Section 4.4.1 while evaluating validity and limits of paleostress inversion methods.

b) Graphical determination of possible fault slips

The possibility of predicting fault slip orientations for specific stress states represents the main outcome of the main Wallace–Bott hypothesis and constitute the background of another class of methods, the so-called direct methods (see e.g. Pascal, 2004 and Chapter 7, Section 7.1 for further details). Fast and relatively effortless exploration of mechanically possible fault slip senses is the main advantage of direct methods. The study can be conducted numerically in very short calculation times, using for example Eq. (3.70), or graphically.

We detail in the following graphical determinations of fault slips for some relevant cases with the primary goal of illustrating the physics behind the Wallace–Bott hypothesis. The method devised here (e.g. Angelier, 1994) is simple but limited to the determination of the range of possible shear vectors or fault slips. More elaborated graphical methods exist and allow for precise determination of shear vectors on planes in function of stress orientations and stress anisotropy, but describing them would be redundant in the context of the present discussion. Interested readers are invited to check e.g. Lisle (1989, 1998), Means (1989), DePaor (1990), Ragan (1990), Fry (1992a), Fleischmann (1992), Ritz (1994) and Shan et al. (2009) for further details.

In order to explain the main principles of the graphical constructions, let us first consider a normal stress regime with shape ratio $\Phi = 0$ (i.e. uniaxial or radial extension, Fig. 4.5A). As noted in Chapter 3 (Section 3.1.3), $\Phi = 0$ corresponds to $\sigma_2 = \sigma_3$ or, conversely, to stress isotropy according to the plane containing both the intermediate and the minimum principal stress, the horizontal plane in the present case. Stress isotropy according to the horizontal plane means that stress magnitude remains constant (i.e. equal to σ_3 in our example) whatever the trend of a randomly selected line in the plane. The latter property reflects the fact that, for uniaxial stress states, the precise orientations

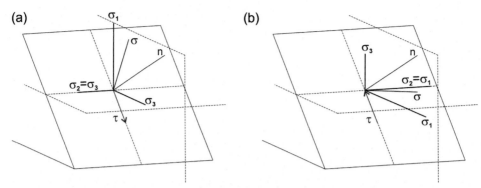

FIG. 4.5 Shear stress (τ) orientation on a fault plane under (A) uniaxial extension ($\Phi = 0$) and (B) uniaxial compression ($\Phi = 1$). For both cases, stress is isotropic according to the horizontal plane and, therefore, the orientations of the horizontal principal axes of stress are arbitrary. The choice of σ_2 parallel to the fault implies that σ_1, σ_3 and the stress vector, σ, belong to the vertical plane containing the normal of the fault, **n**, and consequently that τ is also included in that vertical plane. In conclusion, when $\Phi = 0$ and $\Phi = 1$ the orientation of the shear stress is automatically parallel to the orthogonal projection on the fault plane of σ_1 and σ_3, respectively. The arrows point to the directions of motion of the hanging wall.

of the two principal stresses with equal magnitudes cannot be mathematically determined but need to be arbitrarily located in their corresponding plane. Thus, we are allowed to choose whatever pair of orthogonal axes in the horizontal plane to represent σ_2 and σ_3, provided that they form a right-handed system with σ_1.

We opt for the convenient configuration where σ_2 is parallel to the fault plane and therefore σ_3 oriented along its dip direction (Fig. 4.5A). A principal axis of stress parallel to a plane does not contribute to the stress exerted on that plane (i.e. one of the direction cosines in Eqs (3.48)–(3.50) vanishes). In our example, σ_2 does not contribute to the stress on the fault plane, which implies that the stress vector and, therefore, the shear stress vector are contained in the vertical plane involving σ_3 and σ_1 or, alternatively, involving the normal of the plane under consideration, **n**, and σ_1. Noteworthy, the latter conclusion is independent of the relative orientation of the stress axes with respect to the coordinate system. The particular configuration depicted in Fig. 4.5A was drawn intentionally to facilitate visualisation; however, only the orientation of the plane with respect to the stress axes is relevant here. If σ_1 were oblique to the vertical direction, its geometrical relationship with the shear stress would remain untouched.

In brief, the shear stress vector and (presumably) the slip vector are parallel to the orthogonal projection of σ_1 on the plane for a uniaxial stress state with $\Phi = 0$ (Fig. 4.5A). The same line of reasoning applied to the case of uniaxial compression, where σ_3 is vertical and $\Phi = 1$ (i.e. Fig. 4.5B), leads to a similar conclusion: the slip vector is parallel to the orthogonal projection of σ_3 on the plane for a uniaxial stress state with $\Phi = 1$.

Provided the orientations of the principal axes of stress are known, the interesting implication of the previous findings is that, the slip vectors corresponding to the two extreme values of Φ can be graphically determined, constraining automatically the whole collection of fault slips. Practically, the graphical determination consists in plotting σ_1, σ_3, the fault plane and its corresponding pole on a net and in drawing the planes containing σ_1 and the pole of the fault and σ_3 and the pole, respectively. The intersections between the constructed planes and the fault plane furnish the locations of the shear stresses related to the two extreme values of Φ and limit the cyclographic path of the permitted fault slips on the fault plane (Fig. 4.6). To note, the graphical determination constrains the solution domain but, in the general case, it is not straightforward to associate fault slips within this domain to precise Φ values.

Fig. 4.6A gives examples of graphical determination of fault slips for the three stress regimes, when all principal axes of stress are oblique to the fault. To facilitate the comparison of the results, each stress configuration was created by permutation of two of the stress axes in the following or previously depicted stress regime. The figure illustrates clearly a fact demonstrated mathematically by Bott (1959) and in Chapter 3

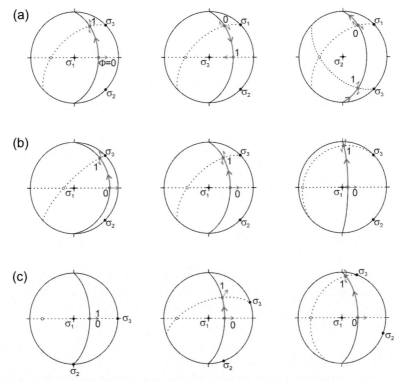

FIG. 4.6 Graphical determination of collections of permitted shear stress vectors (i.e. fault slips). (A) Examples for normal, reverse and strike-slip stress regimes. (B) and (C) depict examples for normal stress states when the relative attitude of the fault plane with respect to three principal axes of stress varies. (B) Variation of the solution domain in function of fault dip. (C) Variation of the solution domain in function of stress orientation. *Open dots* represent fault poles, outwards and inwards *arrows* indicate dominantly normal and reverse fault slips, respectively, and double paired *arrows* dominantly strike slips. The collections of permitted shear stresses are depicted in *grey*, *arrows* point to directions of progressive increase in Φ values from 0 to 1. Projection is on the lower hemisphere of a Schmidt net.

(Section 3.1.4): fault slip orientation depends not only on the orientations of the principal axes of stress but also on the value of the shape ratio. The normal and reverse stress regimes result in similar traces of the solution domain on the net, indicating similar orientations for the fault slips albeit with opposite senses. The prediction of dip slip for $\Phi = 0$, in the case of the normal stress regime, but $\Phi = 1$ for the reverse one, in agreement with Fig. 4.5, implies opposite senses of rotation along the solution paths.

In the example of strike-slip stress regime, fault slips bear systematically dextral strike-slip components, consistent with the relative orientations of the principal axes of stress with respect to the fault plane (swapping σ_1 and σ_3 would result in sinistral components), but alternating reverse and normal dip slip components. The reverse component is maximum for $\Phi = 0$ and decreases as Φ increases until fault slip becomes pure strike-slip (i.e. where the plane of the fault meets the horizontal plane on the net, Fig. 4.6A). If Φ increases further, a normal component appears and reaches its maximum for $\Phi = 1$. This surprising behaviour of changing the sense of the dip slip component is indeed a natural consequence of what we have previously discussed about uniaxial stress states. In simplified but helpful terms, we may imagine that the hanging wall of the fault has the tendency to move according to the direction of 'push' of σ_1 for $\Phi = 0$ (Fig. 4.5), hence the reverse component in the example devised in Fig. 4.6A, and the direction of 'pull' of σ_3 for $\Phi = 1$, thus the normal component.

It is particularly interesting to explore variations of the solution domain in function of relative orientation of the fault with respect to the principal axes of stress. In Fig. 4.6B, the directions of the principal stress axes are kept constant, all axes being oblique to the fault plane, while the dip of the fault plane is progressively augmented. Note that the examples are given for a normal stress regime but the solutions for reverse and strike-slip regimes can easily be deduced from the figure, keeping in mind the relationships presented in Fig. 4.6A. We remark that the higher the dip the longer the cyclographic trace of the range of fault slip solutions, and this effect is particularly well expressed when the fault plane is nearly vertical. The lengthening of the trace of the solution domain is explained by the orthogonal projection of σ_3 on the fault plane, i.e. the fault slip for $\Phi = 1$, which tends to horizontal for steep faults, promoting pronounced strike-slip components.

In Fig. 4.6C, the attitude of the fault is kept constant while the principal axes of stress are progressively rotated counterclockwise in the horizontal plane. In the firstly depicted configuration, σ_3 is parallel to the dip direction of the fault, therefore the two construction planes shown as dashed lines coincide, and the solution domain is represented by a single point corresponding to (normal) pure dip slip. This specific configuration presents obvious similarities with neoformed Andersonian faults but describes the peculiar cases where faults are reactivated while containing the intermediate principal stress. The length of the cyclographic trace of permitted fault slips increases when σ_3 is rotated away from its initial position (Fig. 4.6C) and the smaller the angle between σ_3 and fault strike the longer the trace (i.e. the more pronounced the strike-slip component for $\Phi = 1$). The latter behaviour is once again explained by the orthogonal projection of σ_3, whose horizontal component increases when the minimum principal stress approaches parallelism with the fault plane.

Let us examine the main outcomes of the results presented in Fig. 4.6C. The graphical analysis of the first depicted case shows that only pure dip slip is permitted whatever the value of Φ. The latter conclusion implies that, for this particular case, fault slip is fully resolved by direct methods, assuming that the orientations of the principal axes of stress are known. On the contrary, a dip slip fault furnishes poor constraints on the stress tensor when applying inverse methods, because the stria can correspond to any value of the shape ratio between 0 and 1. A similar conclusion is easily reached for pure dip slip reverse faults and, by simple rotation, one can deduce that this inference holds as well for pure strike-slip faults with vertical planes. In addition, the (normal) dip slip character of the fault would remain independent of Φ if σ_3 were replaced with σ_2 (save for the case $\Phi = 1$, for which shear stress vanishes when σ_3 is parallel to the fault, see Eq. (3.69)). Thus, dip slip faults constrain badly not only the value of Φ but also the orientations of, at least, two principal axes of stress, when inversion of fault slip data is attempted. Faults with long cyclographic traces of the solution domain reduce however the resolution power of direct methods, the range of possible fault slips being relatively large. In turn, they furnish strong constraints to inversion methods in the way that each point along the trace, i.e. the actual stria measured in the field, corresponds to a relatively precise value of the shape ratio for a given set of orientations of the principal axes of stress.

In conclusion, the graphical analyses of fault slips show that, in general, pure dip slip faults and pure strike-slip vertical faults constrain badly paleostress tensors when inversion methods are used. Faults with oblique slips furnish however firmer constraints.

4.1.3 The 'reduced' stress tensor

The graphical constructions devised in the previous section show that the orientation of the shear stress vector depends solely on the orientations of the principal axes of stress and the shape ratio, i.e. the anisotropy of the stress state, a result demonstrated originally by Bott (1959) and in agreement with the analytical development given in Chapter 3, Section 3.1.4. This conclusion simply means that a given fault slip can be explained by an infinity of stress tensors, sharing the same shape ratio and the same orientations for the principal axes of stress but presenting distinct stress magnitudes. One is therefore tempted to find a particular form of the stress tensor, such as it retains exclusively the four parameters shared by all the tensors fitting the observed stria (i.e. Φ and three angles describing the orientations of the stress axes), instead of picking arbitrarily one of the stress tensors honouring the problem. This particular stress tensor, the so-called *reduced stress tensor*, was defined by the French geologist Jacques Angelier (1947–2010), an authority in paleostress studies.

We postpone to the next section the rigorous mathematical treatment of the problem (see Section 4.2.1) and, for now, examine semiqualitatively the expression:

$$T_r = \alpha T + \beta I \qquad (4.1)$$

where T is a stress tensor expressed in the coordinate system of its eigenvectors (see Eq. (3.22)), I is the unit matrix, α and β are scalars and T_r represents the reduced stress tensor. For the sake of clarity, we also postpone the discussion on the signs of the two scalars to the next section and assume positive values for α thereafter.

Multiplication of a stress tensor by a positive scalar is equivalent to multiplying its principal stress magnitudes by this value thus, by virtue of Eqs (3.48)–(3.50), to multiplying the stress vector magnitude and those of its normal and shear components (Fig. 4.7). The operation leaves nevertheless stress orientations and, in particular, the orientation of the shear stress vector untouched (Fig. 4.7B). We remark that this operation results in alteration of both the mean and the differential stresses, that is produces changes in position and size of the corresponding Mohr circle (Fig. 4.7B and Fig. 3.9 for reference). The relative position of the intermediate principal stress between the minimum and the maximum principal stress is however unchanged, meaning that the shape ratio is insensitive to multiplication by a scalar, as its definition as ratio of stress differences automatically implies (see Eq. (3.38)).

Addition of an isotropic component, i.e. the second term to the right side of Eq. (4.1), increments or lowers stress magnitudes by the corresponding amount of pressure (Fig. 4.7C). Note that there is a physical limit to the amount of pressure that can be subtracted (see mathematical treatment in Section 4.2.1.c, if the normal stress becomes

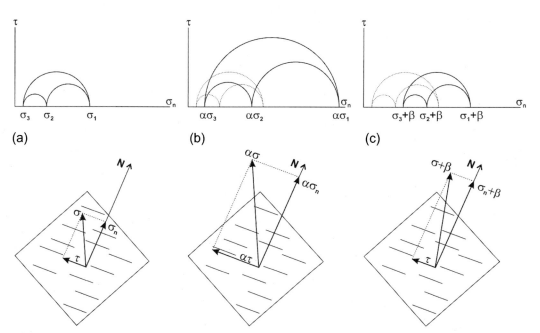

FIG. 4.7 Variation of stress components. (A) Initial stress state. (B) Multiplication of the stress tensor by a positive scalar, α. (C) Addition of an isotropic stress, β. The upper panel shows the different stress states in the Mohr space (in B) and (C) the initial state is depicted as *dashed semicircles*), the lower panel represents a hypothetical striated fault plane and normal (**N**), stress (σ), normal stress (σ_n) and shear stress (τ) vectors. Note that the orientation of the shear stress vector remains constant in all cases.

negative, the fracture opens and shear stress vanishes. Otherwise shear stresses, either in terms of magnitude or orientation, are insensitive to addition or subtraction of pressures, as already discussed in Section 3.1.4. In the Mohr space, the operation corresponds to simple translation of the Mohr circle parallel to the abscissa (see also Fig. 3.18), the size of the circle and the relative position of the intermediate principal stress remaining unaffected. In other words, mean stress varies but stress differences remain constant. Hence, the shape ratio is not affected by the addition of an isotropic stress component.

The previous lines of reasoning convince us that the transformation of the stress tensor given in Eq. (4.1) conserves (1) the orientations of the principal axes of stress, (2) the value of the shape ratio and (3) the orientation of the shear stress vector. The next step consists in choosing α and β in order to find an elegant form for the reduced tensor. Following Angelier (1975), we write:

$$\alpha = \frac{1}{\sigma_1 - \sigma_3} \tag{4.2}$$

$$\text{and } \beta = -\frac{\sigma_3}{\sigma_1 - \sigma_3} \tag{4.3}$$

Substituting the above expressions of α and β into (4.1) gives

$$T_r = \frac{1}{\sigma_1 - \sigma_3} \begin{pmatrix} \sigma_1 & 0 & 0 \\ 0 & \sigma_2 & 0 \\ 0 & 0 & \sigma_3 \end{pmatrix} - \frac{\sigma_3}{\sigma_1 - \sigma_3} \begin{pmatrix} 1 & 0 & 0 \\ 0 & 1 & 0 \\ 0 & 0 & 1 \end{pmatrix} \tag{4.4}$$

and finally

$$T_r = \begin{pmatrix} 1 & 0 & 0 \\ 0 & \Phi & 0 \\ 0 & 0 & 0 \end{pmatrix} \tag{4.5}$$

where Φ is the shape ratio defined in Eq. (3.38).

The reduced stress tensor has the pleasant property of being defined by only four parameters instead of six for the general form of the stress tensor (Chapter 3, Section 3.1.3): Φ and three angles giving the orientations of the three principal axes of stress. It is in that sense that the tensor is 'reduced'. These four parameters represent the geometrical characteristics of the stress tensor whereas the two other parameters, e.g. mean and differential stress, relate to stress magnitudes.

We now determine the expression of the shear stress vector. In agreement with the analytical developments presented in Chapter 3, the shear stress, the stress and the normal stress vectors are, respectively, given by:

$$\vec{\tau}_r = \vec{\sigma}_r - \vec{\sigma}_{nr} \tag{4.6}$$

$$\vec{\sigma}_r = T_r \vec{N} \tag{4.7}$$

$$\text{and } \vec{\sigma}_{nr} = \left(\vec{\sigma}_r \cdot \vec{N} \right) \vec{N} \tag{4.8}$$

where the symbols keep the same meanings as in Chapter 3 and the index 'r' recalls that the vectors are derived from the reduced stress tensor.

After developing (4.8) we find:

$$\vec{\sigma}_r = \begin{pmatrix} 1 & 0 & 0 \\ 0 & \Phi & 0 \\ 0 & 0 & 0 \end{pmatrix} \begin{pmatrix} l \\ m \\ n \end{pmatrix} \tag{4.9}$$

$$\vec{\sigma}_r = \begin{pmatrix} l \\ m\Phi \\ 0 \end{pmatrix} \tag{4.10}$$

l, m and n being the direction cosines of the normal vector as defined in Chapter 3.

Eq. (4.8) gives:

$$\vec{\sigma}_{nr} = (l^2 + m^2\Phi) \begin{pmatrix} l \\ m \\ n \end{pmatrix} \tag{4.11}$$

and combining (4.6), (4.10) and (4.11), the 'reduced' shear stress vector is:

$$\vec{\tau}_r = \begin{pmatrix} l(1 - l^2 - m^2\,\Phi) \\ m(-l^2 + (1 - m^2)\,\Phi) \\ -n\,(l^2 + m^2\,\Phi) \end{pmatrix} \tag{4.12}$$

Using Eq. (3.43), $1 - l^2$ and $1 - m^2$ are replaced with $m^2 + n^2$ and $l^2 + n^2$, respectively, in (4.12).

We finally find:

$$\vec{\tau}_r = \begin{pmatrix} l\,(n^2 + m^2\,(1 - \Phi)) \\ m\,(n^2\Phi - l^2\,(1 - \Phi)) \\ -n\,(m^2\Phi + l^2) \end{pmatrix} \tag{4.13}$$

which is the general expression of the shear stress vector given by Eq. (3.70) normalised by the differential stress. As consequence of e.g. Eq. (3.81), the normalised and signed length of the 'reduced' shear stress vector ranges between -0.5 and $+0.5$. Expectedly, Eq. (4.13) carries information about orientation of the shear stress vector (and its relative length with respect to differential stress) but knowledge about stress magnitude is lost. The 'reduced' shear stress furnishes therefore a level of information equivalent to what can be extracted from the measurement of fault striae in the field. Consequently, reduced stress tensors are particularly adapted to handle stress-slip relationships when further information about stress magnitudes cannot be easily gathered from fieldwork as it is commonly the case.

4.2 Numerical inversion of fault slip data

4.2.1 Theoretical background

a) The mechanical model

The problem of inversion of fault slip data was for the first time addressed by Arthaud (1969). Arthaud's method was based on graphical constructions of the *plane of motion* for each fault (i.e. the plane containing the stria and the normal of the fault). The intersection of these planes was assumed to correspond to either the minimum or the maximum principal axis of strain. Arthaud judged that interactions between faults precluded the determination of principal axes of stress. In case of fault interaction, the stress state was indeed locally disturbed and the stria did not correspond any longer to the 'global' stress state acting on the studied rock volume. The method was reminiscent of the approaches advanced by Turner and Weiss (1963) for the stress analysis of e.g. calcite twins (see Chapter 6, Section 6.3.1) as it relied on the inherent symmetry of the system. Carey and Brunier (1974) pointed out that Arthaud's method presented the serious drawback of being limited to the resolution of uniaxial (either strain or stress) states. The latter limitation is a direct consequence of the main hypothesis of the method, which implies that the stria lies along the orthogonal projection of one of the two extreme principal axes (see discussion in Section 4.1.2).

Arthaud's work was nevertheless instrumental for the setting up of the foundations of fault slip inversion methods for it explored in unprecedented detail the physical conditions pertaining to their validity. Following Bott (1959), Arthaud assumed that: (1) deformation in the brittle domain was mainly accommodated by finite displacements along preexisting planar discontinuities, e.g. beddings, joints or metamorphic foliations, and (2) deformation of the faulted blocks was chiefly elastic and consequently negligible. Fault nucleation was supposed to occur only in the seldom cases of rocks practically void of discontinuities. However, neoformed faults could be accounted for by the model as newly discontinuities formed during the course of the tectonic event but prior to their latest displacements. As such volumetric strain in the brittle domain of a given volume of rock was thought of as the sum of finite and small displacements along numerous discontinuities within the considered volume of rock divided by the typical dimensions of the volume.

Carey and Brunier (1974) adopted the main Wallace–Bott hypothesis (see Section 4.1.2) and Arthaud's assumptions for the development of the very first numerical method of paleostress determination by means of inversion of fault slip data[d]. In order to ensure homogenous deformation conditions, they advanced three additional assumptions: fault slips remain modest, with respect to the dimensions of the studied volume of rock, and independent from one another and no torques act on the material (see also Carey, 1976, 1979). Together with the implicit hypothesis that the measured fault slips reflect

---

[d]In the mid 70s, three scientists were actively investigating the topic in parallel: Evelyne Carey and Jacques Angelier, in France, and Oleg Gushchenko in Russia. Gushchenko published in Russian (Gushchenko, 1975, 1979) and unfortunately his contribution remained practically unknown in western countries until the 90s.

a single and uniform paleostress state, the additional assumptions of Arthaud and Carey are substantially the 'Wallace–Bott' hypotheses previously listed under (2) to (6) in Section 4.1.2. Carey (1976) pointed out that for such a simplified mechanical model the principal axes of stress were implicitly parallel to the principal axes of strain. Furthermore, she noted that the model could be validated only a posteriori and in front of the consistency and usefulness of its results.

b) Mathematical statement of the problem

Keeping in mind the simplifying assumptions of the background mechanical model described above, the problem consists in finding a reduced stress tensor that explains the striae measured in the field or, conversely, in determining a reduced stress tensor resulting in shear stress vectors parallel to the measured striae. The mathematical treatment proposed here follows mainly the original analysis of Carey (1976).

Let us define the vector $\vec{u}_i$ orthogonal to both the normal of the studied fault plane, $\vec{n}_i$, and the stria, $\vec{s}_i$, such as the three vectors form a right-hand rule orthonormal set (Fig. 4.8):

$$\vec{u}_i = \vec{s}_i \times \vec{n}_i \tag{4.14}$$

$$u_i = s_i = n_i = 1 \tag{4.15}$$

where $\times$ represents vector or cross product.

According to the main Wallace–Bott hypothesis $\vec{s}_i$ and $\vec{\tau}_i$ are parallel; therefore, the problem to be solved may be written:

$$\vec{\sigma}_i \cdot \vec{u}_i = 0 \tag{4.16}$$

$$\text{or } \vec{\tau}_i \cdot \vec{u}_i = 0 \tag{4.17}$$

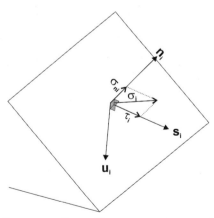

FIG. 4.8 Fault plane with normal, $\vec{n}_i$, fault slip, $\vec{s}_i$, and $\vec{u}_i$ vectors forming an orthonormal basis. Stress, $\vec{\sigma}_i$, normal stress, $\vec{\sigma}_{ni}$, and shear stress, $\vec{\tau}_i$, vectors are also depicted. Note that the sense of both the shear stress and the stria correspond to the direction of displacement of the hanging wall.

Each one of the two conditions involves the input geometrical parameters of the fault plane and of the stria and the unknowns of the stress tensor. It is thus reasonable to anticipate that the latter can be derived from the former. To note, conditions (4.16) and (4.17) ensure parallel shear stress and stria vectors but these may be of same or opposite sense, requiring an additional condition to better constrain the solution. The primary aim of this subsection is, nevertheless, the demonstration of the mathematical validity of fault slip inversion methods. Modifications and refinements are devised in the following subsections.

c) Exploration of the space of solutions

We consider the stress tensor, T_0, as solution of Eqs (4.16) and (4.17). Let us write the stress tensor, T, such as:

$$T = \alpha T_0 + \beta I \tag{4.18}$$

where α and β are real numbers and I indicates the unit matrix, as usual.

The stress vectors associated with T_0 and T, and acting on the plane indexed by i are respectively:

$$\vec{\sigma}_{0i} = T_0 \vec{n}_i \tag{4.19}$$

$$\text{and } \vec{\sigma}_i = T \vec{n}_i \tag{4.20}$$

Combining (4.18), (4.19) and (4.20):

$$\vec{\sigma}_i = (\alpha T_0 + \beta I) \vec{n}_i \tag{4.21}$$

$$\text{and finally } \vec{\sigma}_i = \alpha \vec{\sigma}_{0i} + \beta \vec{n}_i \tag{4.22}$$

Therefore the scalar product of $\vec{\sigma}_i$ and $\vec{u}_i$ is given by:

$$\vec{\sigma}_i \cdot \vec{u}_i = \alpha \vec{\sigma}_{0i} \cdot \vec{u}_i + \beta \vec{n}_i \cdot \vec{u}_i \tag{4.23}$$

Recalling that $\vec{\sigma}_{0i}$ and $\vec{n}_i$ are both orthogonal to $\vec{u}_i$ we find:

$$\vec{\sigma}_i . \vec{u}_i = 0 \tag{4.24}$$

We now write the expressions of the shear stress vector associated with the stress tensor T:

$$\vec{\tau}_i = \vec{\sigma}_i - \left(\vec{\sigma}_i . \vec{n}_i \right) \vec{n}_i \tag{4.25}$$

The scalar product of $\vec{\tau}_i$ and $\vec{u}_i$ is:

$$\vec{\tau}_i \cdot \vec{u}_i = \vec{\sigma}_i \cdot \vec{u}_i - \left(\vec{\sigma}_i \cdot \vec{n}_i \right) \vec{n}_i \cdot \vec{u}_i \tag{4.26}$$

$$\text{and thus } \vec{\tau}_i . \vec{u}_i = 0 \tag{4.27}$$

In summary, if a stress tensor T_0 is solution of the problem any tensor T defined by Eq. (4.18) is also solution. The analysis is also a rigorous demonstration that reduced stress tensors are valid solutions of the inversion problem, as previously suggested by the qualitative considerations detailed in Section 4.1.3.

The mathematical analysis demonstrates that all tensors satisfying Eq. (4.18) are solutions of the problem but does not prove that the equation involves all possible solutions.

Let us express T in the generalised form:

$$T = \alpha T_0 + \sum_{j=1}^{M} \alpha_j T_j + \beta I \tag{4.28}$$

where j and M are positive integers, all other symbols keeping their respective meanings as indicated previously.

For each plane referred by i, each stress vector $\vec{\sigma}_{ji}$ derives from one tensor T_j and, in addition, application of the isotropic stress βI results in a stress vector parallel to the normal of the plane $\vec{n}_i$. Assuming that T is solution of the problem, all the orthogonal projections on the fault plane of the stress vectors $\vec{\sigma}_{ji}$ should be parallel to the stria $\vec{s}_i$, that is the collection of $\vec{\sigma}_{ji}$ (and also $\vec{\sigma}_{0i}$) belong to the plane of motion containing $\vec{n}_i$ and $\vec{s}_i$.

However, no more than two vectors belonging to the same 2D plane can be independent. Therefore, it is always possible to write each $\vec{\sigma}_{ji}$ in function of $\vec{\sigma}_{0i}$ and $\vec{n}_i$:

$$\vec{\sigma}_{ji} = g_j \vec{\sigma}_{0i} + h_j \vec{n}_i \tag{4.29}$$

where g_j and h_j are real numbers.

By definition Eq. (4.29) derives from:

$$T_j = g_j T_0 + h_j I \tag{4.30}$$

Combining (4.28) with (4.30) we find:

$$T = \left(\alpha + \sum_{j=1}^{M} \alpha_j g_j \right) T_0 + \left(\beta + \sum_{j=1}^{M} h_j \right) I \tag{4.31}$$

which is strictly analogous to Eq. (4.18).

In conclusion, we have demonstrated that any tensor solution of the problem can always be written in the form given by Eq. (4.18) or, conversely, that Eq. (4.18) represents all the possible mathematical solutions to the fault slip inversion problem.

However, not all mathematical solutions are relevant to the physical problem.

Let us first consider the particular cases for which $\alpha = 0$:

$$T = \beta I \tag{4.32}$$

$$\vec{\sigma}_i = \beta \vec{n}_i \tag{4.33}$$

$$\text{and} \ \ \vec{\tau}_i = 0 \tag{4.34}$$

The result appears to be somewhat trivial as we have already seen that shear stress vanishes in the case of isotropic stress states. Nevertheless, it is worth noting that, although no shear is predicted, the particular cases given by Eq. (4.32) are solutions of the mathematical problem but, definitively, irrelevant to the physical problem.

Cases where stresses normal to the fault plane are negative (i.e. tensional) are also irrelevant to the physical problem. For these specific cases, negative normal stresses result in opening of the fault plane, contact loss between the faulted blocks and, therefore, absence of friction stria. In more detail, if $\alpha \neq 0$ shear displacements are predicted (and sometimes observed) anyway, but the fracture adopts the kinematics of a hybrid fracture and no friction marks are to be expected. The fault slip vector can potentially be determined from the displacement of preexisting markers in the rock; however, this latter possibility is rarely met in the field.

In order to determine the solutions relevant to the physical problem, we combine Eqs (3.53) and (4.22) and write the expression of the normal stress vector:

$$\sigma_{ni} = \alpha \sigma_{n0i} + \beta \tag{4.35}$$

where $\sigma_{n0i} \geq 0$ and $\alpha \neq 0$.

According to Eq. (4.35), if α and β are both positive then σ_{ni} is always positive and all the corresponding stress tensors represent relevant physical solutions. Alternatively, if α and β are both negative, σ_{ni} is always negative and none of the stress tensors is relevant to the physical problem.

For the remaining cases let us examine:

$$\sigma_{ni} \geq 0 \tag{4.36}$$

$$\text{or alternatively} \quad \alpha \sigma_{n0i} + \beta \geq 0 \tag{4.37}$$

Condition (4.37) is met:

$$\text{if} \quad \sigma_{n0i} \leq \frac{-\beta}{\alpha} \quad \text{for } \alpha < 0 \text{ and } \beta > 0 \tag{4.38}$$

$$\text{and if} \quad \sigma_{n0i} \geq \frac{-\beta}{\alpha} \quad \text{for } \alpha > 0 \text{ and } \beta < 0 \tag{4.39}$$

Before closing the exploration of the solutions of the fault slip inversion problem, let us express T_0 in function of its six components:

$$T_0 = \begin{pmatrix} x_{01} & x_{04} & x_{05} \\ x_{04} & x_{02} & x_{06} \\ x_{05} & x_{06} & x_{03} \end{pmatrix} \tag{4.40}$$

By definition (see Section 3.1.3) the trace of T_0 is given by:

$$\text{Trace}(T_0) = x_{01} + x_{02} + x_{03} \tag{4.41}$$

According to Eq. (4.18), the trace of T is therefore:

$$\text{Trace}(T) = \alpha \, \text{Trace}(T_0) + 3\beta \tag{4.42}$$

Setting $\alpha = 1$ and $\beta = -\text{Trace}(T_0)/3$, we find $\text{Trace}(T) = 0$. Thus, T takes the special form of a deviator (see Eq. (3.37)).

In conclusion, it is always possible to find a deviator in the collection of solutions defined by Eq. (4.18).

Furthermore, all deviators solution of the problem are linearly dependent on each other. To demonstrate the latter statement, let us define two deviators solution of the problem T_{D0} and T_{D1} such as:

$$T_{D0} = T_0 - \frac{1}{3}\text{Trace}(T_0)I \tag{4.43}$$

$$T_{D1} = T_1 - \frac{1}{3}\text{Trace}(T_1)I \tag{4.44}$$

with

$$T_1 = \alpha T_0 + \beta I \tag{4.45}$$

$$\text{Trace}(T_1) = \alpha\,\text{Trace}(T_0) + 3\beta \tag{4.46}$$

Combining Eqs (4.44)–(4.46) we find:

$$T_{D1} = \alpha T_0 + \beta I - \frac{\alpha}{3}\text{Trace}(T_0)I - \beta I \tag{4.47}$$

$$T_{D1} = \alpha\left(T_0 - \frac{\text{Trace}(T_0)}{3}I\right) \tag{4.48}$$

and according to Eq. (4.43)

$$T_{D1} = \alpha T_{D0} \tag{4.49}$$

d) Main properties of the family of solutions

Let us consider one of the eigenvalues of T_0, λ_0, and its related eigenvector $\vec{U}_0$. By definition:

$$T_0\vec{U}_0 = \lambda_0\vec{U}_0 \tag{4.50}$$

According to (4.18), application of T on $\vec{U}_0$ gives:

$$T\vec{U}_0 = \alpha T_0\vec{U}_0 + \beta I\vec{U}_0 \tag{4.51}$$

$$\text{and finally} \quad T\vec{U}_0 = (\alpha\lambda_0 + \beta)\vec{U}_0 \tag{4.52}$$

Eq. (4.52) demonstrates that all the tensors solution of the problem share the same eigenvectors (i.e. principal axes of stress) but differ from each other in terms of eigenvalues (i.e. principal stress magnitudes). The result confirms what has already been inferred in Section 4.1.3 based on qualitative arguments.

We examine now the three eigenvalues of T_0, λ_1, λ_2 and λ_3 ordered such as $\lambda_3 \leq \lambda_2 \leq \lambda_1$ and $\lambda_1 \neq \lambda_3$. Note that the symmetry of the tensor ensures that the three eigenvalues exist and are real numbers. The three eigenvalues of T are $\alpha\lambda_1 + \beta$, $\alpha\lambda_2 + \beta$ and $\alpha\lambda_3 + \beta$, and

$$\text{if } \alpha > 0 \quad \alpha\lambda_3 + \beta \leq \alpha\lambda_2 + \beta \leq \alpha\lambda_1 + \beta \tag{4.53}$$

$$\text{if } \alpha < 0 \quad \alpha\lambda_3 + \beta \geq \alpha\lambda_2 + \beta \geq \alpha\lambda_1 + \beta \tag{4.54}$$

These simple operations demonstrate that the eigenvector corresponding to the intermediate eigenvalue remains the same for all tensors solution of the problem.

In addition, simple manipulations of Eq. (4.53), or alternatively (4.54), lead to the conclusion that all the tensors solution of the problem (save the tensors representing isotropic stress states) share a common relationship between their respective eigenvalues, which defines the degree of stress anisotropy. The latter relationship may take different mathematical forms, e.g.:

$$0 \leq \frac{\lambda_2 - \lambda_3}{\lambda_1 - \lambda_3} \leq 1 \tag{4.55}$$

which corresponds to the shape ratio Φ traditionally used by Angelier (1975) and many others,

$$\text{or } 0 \leq \frac{\lambda_2 - \lambda_1}{\lambda_3 - \lambda_1} \leq 1 \tag{4.56}$$

as in e.g. Gephart (1990a).

e) Determination of the solutions

For the sake of the demonstration, we need to approach the mathematical problem as a geometrical one in the 6D space corresponding to the six unknowns of the stress tensor. Thus, a stress tensor is analogous to a point referred by six values.

That is the tensor (solution or not of the problem)

$$T_G = \begin{pmatrix} x_1 & x_4 & x_5 \\ x_4 & x_2 & x_6 \\ x_5 & x_6 & x_3 \end{pmatrix} \tag{4.57}$$

is represented by point P whose coordinates are $(x_1, x_2, x_3, x_4, x_5, x_6)$ and referred by the vector defined from the origin to point P:

$$\vec{V}_P = \begin{pmatrix} x_1 \\ x_2 \\ x_3 \\ x_4 \\ x_5 \\ x_6 \end{pmatrix} \tag{4.58}$$

Similarly, the equation of solutions, $T = \alpha T_0 + \beta I$, represents lines and all the solutions are contained in the plane defined by these lines in the 6D space.

By definition, line L, characterising the collection of isotropic stress tensors, $T=\beta I$, belongs to the plane of solutions. According to the convention given above, L has for unit vector:

$$\vec{V}=\begin{pmatrix} \dfrac{1}{\sqrt{3}} \\ \dfrac{1}{\sqrt{3}} \\ \dfrac{1}{\sqrt{3}} \\ 0 \\ 0 \\ 0 \end{pmatrix} \tag{4.59}$$

Let us write the scalar product of $\vec{V}$ and $\vec{V}_P$:

$$\vec{V}\cdot\vec{V}_P=\begin{pmatrix} 1/\sqrt{3} \\ 1/\sqrt{3} \\ 1/\sqrt{3} \\ 0 \\ 0 \\ 0 \end{pmatrix}\begin{pmatrix} x_1 \\ x_2 \\ x_3 \\ x_4 \\ x_5 \\ x_6 \end{pmatrix} \tag{4.60}$$

$$\vec{V}\cdot\vec{V}_P=\frac{1}{\sqrt{3}}(x_1+x_2+x_3) \tag{4.61}$$

One interesting outcome of Eq. (4.61) is that $\vec{V}\cdot\vec{V}_P=0$ only if $x_1+x_2+x_3=0$, which means only for stress deviators. Therefore, L is orthogonal to the hyperplane (i.e. 5D plane in the present case), Π, containing all the deviators of the 6D space. We have previously shown that the space of solutions always contains deviators; consequently, the intersection between Π and the solution plane represents the collection of deviators solution of the problem. Moreover, Eq. (4.49) demonstrates that this intersection is a line and that this line is unique. Determining the intersection line, Δ, is equivalent to finding the solution plane, i.e. to solving the inversion problem. We thus devote the rest of the mathematical treatment to these aims.

Let us come back to the basic statement of the problem as it was previously formalised in Eq. (4.27). Independently of the adopted mathematical approach, the problem consists in finding the solutions that satisfy the set of conditions for N fault slip measurements:

$$\vec{\tau}_i\cdot\vec{u}_i=0 \quad \text{with } 1\leq i\leq N \tag{4.62}$$

We write the respective expressions of the two vectors $\vec{n}_i$ and $\vec{u}_i$:

$$\vec{n}_i=\begin{pmatrix} l_i \\ m_i \\ n_i \end{pmatrix} \tag{4.63}$$

$$\vec{u}_i=\begin{pmatrix} x_{ui} \\ y_{ui} \\ z_{ui} \end{pmatrix} \tag{4.64}$$

and calculate the stress vector for each plane using the form of the stress tensor given in Eq. (4.57):

$$\vec{\sigma}_i = \begin{pmatrix} x_1 & x_4 & x_5 \\ x_4 & x_2 & x_6 \\ x_5 & x_6 & x_3 \end{pmatrix} \vec{n}_i \tag{4.65}$$

$$\vec{\sigma}_i = \begin{pmatrix} x_1 l_i + x_4 m_i + x_5 n_i \\ x_4 l_i + x_2 m_i + x_6 n_i \\ x_5 l_i + x_6 m_i + x_3 n_i \end{pmatrix} \tag{4.66}$$

As usual the shear stress vector is:

$$\vec{\tau}_i = \vec{\sigma}_i - \left(\vec{\sigma}_i \cdot \vec{n}_i \right) \vec{n}_i \tag{4.67}$$

where the scalar product of $\vec{\sigma}_i$ by $\vec{n}_i$ takes the form:

$$\vec{\sigma}_i \cdot \vec{n}_i = l_i^2 x_1 + m_i^2 x_2 + n_i^2 x_3 + 2(l_i m_i x_4 + l_i n_i x_5 + n_i m_i x_6) \tag{4.68}$$

After developing and rearranging Eq. (4.67), the three components of the shear stress vector for plane i are:

$$\begin{cases} \tau_{ix} = l_i\left(1 - l_i^2\right)x_1 - l_i m_i^2 x_2 - l_i n_i^2 x_3 + m_i\left(1 - 2l_i^2\right)x_4 + n_i\left(1 - 2l_i^2\right)x_5 - 2l_i m_i n_i x_6 \\ \tau_{iy} = -l_i^2 m_i x_1 + m_i\left(1 - 2m_i^2\right)x_2 - n_i^2 m_i x_3 + l_i\left(1 - 2m_i^2\right)x_4 - 2l_i m_i n_i x_5 + n_i\left(1 - 2m_i^2\right)x_6 \\ \tau_{iz} = -l_i^2 m_i x_1 - n_i m_i^2 x_2 + n_i\left(1 - n_i^2\right)x_3 - 2l_i m_i n_i x_4 + l_i\left(1 - 2n_i^2\right)x_5 + m_i\left(1 - 2n_i^2\right)x_6 \end{cases} \tag{4.69}$$

Combining Eqs (4.62), (4.64) and (4.69) we find:

$$\vec{\tau}_i \cdot \vec{u}_i = a_{i1}x_1 + a_{i2}x_2 + a_{i3}x_3 + a_{i4}x_4 + a_{i5}x_5 + a_{i6}x_6 = 0 \quad \text{with } 1 \le i \le N \tag{4.70}$$

where the a_{ij} coefficients are functions of the coordinates of the normal of the plane and of the stria, i.e. of the quantities measured in the field.

Let us write the vector $\vec{W}_i$ attached to each fault plane i and whose components are the a_{ij} coefficients:

$$\vec{W}_i = \begin{pmatrix} a_{i1} \\ a_{i2} \\ a_{i3} \\ a_{i4} \\ a_{i5} \\ a_{i6} \end{pmatrix} \tag{4.71}$$

Examining (4.70), we conclude that each individual equation represents the scalar product of $\vec{W}_i$ by $\vec{V}_P$, the vector associated with the stress tensor to be found:

$$\vec{W}_i \cdot \vec{V}_P = \begin{pmatrix} a_{i1} \\ a_{i2} \\ a_{i3} \\ a_{i4} \\ a_{i5} \\ a_{i6} \end{pmatrix} \begin{pmatrix} x_1 \\ x_2 \\ x_3 \\ x_4 \\ x_5 \\ x_6 \end{pmatrix} \tag{4.72}$$

Assuming that the stress tensor, whose components are the collection of x_i, is solution of the problem, each equation given in (4.70) should be equal to zero. Therefore:

$$\vec{W}_i \cdot \vec{V}_P = 0 \ \text{ for } 1 \leq i \leq N \tag{4.73}$$

In other words, each equation of the linear system (4.70) represents a hyperplane containing the origin and whose normal is parallel to $\vec{W}_i$ in the 6D space. If the plane of solutions exists, it is a 2D space belonging to the N hyperplanes.

It follows from our discussion in Section 4.1.3 that fault slip inversion methods can only resolve four parameters of the stress tensor. Thus, one expects four independent fault slip data to be sufficient to solve the problem. We demonstrate now the validity of this latter inference and use the demonstration to illustrate the steps of the mathematical reasoning, which is straightforward to generalise to N fault slip data.

Let us consider four hyperplanes as defined by Eq. (4.70) and their respective normals $\vec{W}_1$, $\vec{W}_2$, $\vec{W}_3$ and $\vec{W}_4$. If the space of solutions is shared by the four hyperplanes then it is automatically orthogonal to $\vec{W}_1$, $\vec{W}_2$, $\vec{W}_3$ and $\vec{W}_4$. We have previously demonstrated that the solution plane contains the line L, representing the collection of isotropic stress states, and its vector $\vec{V}$ (Eq. (4.59)).

We introduce the space H generated by the five vectors $\vec{V}$, $\vec{W}_1$, $\vec{W}_2$, $\vec{W}_3$ and $\vec{W}_4$. By definition, H is a hyperplane of the 6D space, whose unique normal $\vec{W}_H$ can be defined and satisfies:

$$\begin{cases} \vec{W}_H \cdot \vec{W}_1 = 0 \\ \vec{W}_H \cdot \vec{W}_2 = 0 \\ \vec{W}_H \cdot \vec{W}_3 = 0 \\ \vec{W}_H \cdot \vec{W}_4 = 0 \end{cases} \tag{4.74}$$

Thus, $\vec{W}_H$ belongs to the solution plane. Furthermore, $\vec{V}$ belonging by construction to H:

$$\vec{W}_H \cdot \vec{V} = 0 \tag{4.75}$$

Therefore, according to Eq. (4.61), $\vec{W}_H$ belongs to Π, the hyperplane containing all the stress deviators. In conclusion, $\vec{W}_H$ belongs to the intersection line between the solution plane and Π, that is, $\vec{W}_H$ is a vector of the line Δ, the line we were searching to determine in order to solve the fault slip inversion problem.

Summarising, we have demonstrated the existence and the mathematical properties of the solutions of the fault slip inversion problem, and proved that the solutions can be rigorously determined. We pursue our discussion with the practical solving of the problem.

4.2.2 Solving the numerical problem

a) Explicit forms of the equations to be solved

Independently of the form selected for the reduced stress tensor and of the adopted mathematical strategy (see next paragraph), all fault slip inversion methods require computation of the orientation of the maximum shear stress acting on the fault plane (save the ones proposed by Carey and Brunier, 1974 and Angelier, 1975, which consider the orientation of the stress vector itself). Thus, most methods involve the same fundamental equations, though written and arranged in different manners.

In order to solve the inverse problem, that is, in order to compare the orientation of the shear stress vector with the one of the stria measured in the field, the components of the reduced stress tensor need to be expressed in the physical coordinate system:

$$T_{xyz} = \Omega T_R \Omega^T \tag{4.76}$$

where T_{xyz} is the reduced stress tensor expressed in the physical coordinate system, Ω the *rotation matrix* and Ω^T its transpose. By definition, the rotation matrix, Ω, transforms coordinates in the reference system of the eigenvectors of the tensor to coordinates in the physical coordinate system. Thus, the rotation matrix may be written:

$$\Omega = \begin{pmatrix} L_1 & L_2 & L_3 \\ M_1 & M_2 & M_3 \\ N_1 & N_2 & N_3 \end{pmatrix} \tag{4.77}$$

where its components represent the director cosines of three principal axes of stress such as:

$$\hat{\sigma}_i = \begin{pmatrix} L_i \\ M_i \\ N_i \end{pmatrix} \quad i \in [1;3] \tag{4.78}$$

the circumflex indicating unit vectors.

Combining Eqs (4.76) and (4.77), we find:

$$T_{xyz} = \begin{pmatrix} L_1 & L_2 & L_3 \\ M_1 & M_2 & M_3 \\ N_1 & N_2 & N_3 \end{pmatrix} \begin{pmatrix} \lambda_1 & 0 & 0 \\ 0 & \lambda_2 & 0 \\ 0 & 0 & \lambda_3 \end{pmatrix} \begin{pmatrix} L_1 & M_1 & N_1 \\ L_2 & M_2 & N_2 \\ L_3 & M_3 & N_3 \end{pmatrix} \tag{4.79}$$

and

$$T_{xyz} = \begin{pmatrix} L_1^2\lambda_1 + L_2^2\lambda_2 + L_3^2\lambda_3 & L_1M_1\lambda_1 + L_2M_2\lambda_2 + L_3M_3\lambda_3 & L_1N_1\lambda_1 + L_2N_2\lambda_2 + L_3N_3\lambda_3 \\ L_1M_1\lambda_1 + L_2M_2\lambda_2 + L_3M_3\lambda_3 & M_1^2\lambda_1 + M_2^2\lambda_2 + M_3^2\lambda_3 & M_1N_1\lambda_1 + M_2N_2\lambda_2 + M_3N_3\lambda_3 \\ L_1N_1\lambda_1 + L_2N_2\lambda_2 + L_3N_3\lambda_3 & M_1N_1\lambda_1 + M_2N_2\lambda_2 + M_3N_3\lambda_3 & N_1^2\lambda_1 + N_2^2\lambda_2 + N_3^2\lambda_3 \end{pmatrix} \tag{4.80}$$

with λ_i representing the three eigenvalues of the reduced stress tensor and $i \in [1;3]$.

Computing the maximum shear stress vector on a given fault plane from the above expression of the stress tensor leads substantially to the system of Eqs (4.69), where the six unknowns x_1 to x_6 are simple functions of the nine cosine directors and of the three eigenvalues of the reduced stress tensor. Thus, Eq. (4.70) becomes:

$$a_{i1}\left[\sum_{j=1}^{3}L_j^2\lambda_j\right] + a_{i2}\left[\sum_{j=1}^{3}M_j^2\lambda_j\right] + a_{i3}\left[\sum_{j=1}^{3}N_j^2\lambda_j\right]$$

$$+ a_{i4}\left[\sum_{j=1}^{3}L_jM_j\lambda_j\right] + a_{i5}\left[\sum_{j=1}^{3}L_jN_j\lambda_j\right] + a_{i6}\left[\sum_{j=1}^{3}M_jN_j\lambda_j\right] = 0 \ \text{ with } 1 \le i \le N \tag{4.81}$$

remembering that N is the total number of fault slip data and that the a_{ij} coefficients represent the input to the problem, i.e. depend on the attitudes of the fault planes and the orientations of their corresponding slips.

We have demonstrated, however, that the three eigenvalues can be reduced to one single value, the shape ratio (Section 4.1.3). In addition, the three eigenvectors of the stress tensor (i.e. the three normalised principal axes of stress) form an orthonormal base, thus:

$$L_i^2 + M_i^2 + N_i^2 = 1 \ \ i \in [1; 3] \tag{4.82}$$

and

$$L_iL_j + M_iM_j + N_iN_j = 0 \ \ i, j \in [1; 3] \ \text{ with } i \ne j \tag{4.83}$$

In brief, six independent equations constrain the nine director cosines, leaving only three unknown angles. In spite of the apparent complexity of Eqs (4.80) and (4.81), only four unknowns remain to be solved, as it was demonstrated in the previous section.

As pointed out in Section 4.2.1, the solving of the fault slip inversion problem requires, theoretically, only four independent equations out of the system of Eqs (4.81), that is, four distinct fault slip data. The problem has therefore a simple analytical solution, in theory. Practically, natural data contain measurement uncertainties and natural noise (e.g. minute displacements of the rocks caused by surface processes). Thus, the solving of the fault slip inversion problem demands numerical strategies and relatively large number of data, as it is the general case when dealing with physical measurements.

As a final remark, the orientations of the stress axes (and of the normal of the plane) are, in practice, more conveniently expressed in terms of trend, t, and plunge, p, in the equations to be solved, that is Eq. (4.78) is replaced by:

$$\hat{\sigma}_i = \begin{pmatrix} \cos p_i \cos t_i \\ \cos p_i \sin t_i \\ \sin p_i \end{pmatrix} \ \ i \in [1; 3] \tag{4.84}$$

b) Choice of the reduced stress tensor and computation strategies

The mathematical treatment detailed in Section 4.2.1 shows that there is an infinity of reduced stress tensors that satisfy the inversion problem. Although the focus of the demonstration was on particular solutions, i.e. deviators, other admissible types of tensors may be constructed according to Eq. (4.18).

The choice of the reduced stress tensor is chiefly dependent on the strategy adopted to solve numerically the inversion problem. Let us examine three different expressions of the reduced stress tensor, T_R, representing the most traditional forms implemented in fault slip inversion codes:

$$T_R = \begin{pmatrix} 1 & 0 & 0 \\ 0 & \Phi & 0 \\ 0 & 0 & 0 \end{pmatrix} \tag{4.85}$$

$$T_R = \begin{pmatrix} 2-\Phi & 0 & 0 \\ 0 & 2\Phi-1 & 0 \\ 0 & 0 & -(\Phi+1) \end{pmatrix} \tag{4.86}$$

and

$$T_R = \begin{pmatrix} \cos\psi & d & e \\ d & \cos\left(\psi+\dfrac{2\pi}{3}\right) & f \\ e & f & \cos\left(\psi+\dfrac{4\pi}{3}\right) \end{pmatrix} \tag{4.87}$$

The reduced stress tensor in Eq. (4.85) is the classical one introduced by Angelier (1975), already discussed in Section 4.1.3, and extensively used in fault slip inversion methods (e.g. Angelier, 1984; Michael, 1984; Yamaji, 2000; Delvaux and Sperner, 2003; Žalohar and Vrabec, 2007; Sasvári and Baharev, 2014). This tensor is not a deviator but renders the shape ratio explicit and it is particularly convenient to illustrate the physical meaning of the results of the inversion.

Eq. (4.86) shows an alternative form chosen by Etchecopar et al. (1981) and Etchecopar (1984), and later used in calcite twin inversion (Tourneret and Laurent, 1990 and Chapter 6). The shape ratio remains explicit but this particular tensor is a deviator, and consequently furnishes an additional constraint to the problem, that is, the sum of its diagonal components, or trace, is equal to zero (see Section 3.1.3, Eq. (3.37)).

The third expression given by Eq. (4.87) is also a deviator, used originally in Angelier and Goguel (1979) and later in e.g. Angelier et al. (1982), Angelier (1990), Will and Powell (1991), Xu (2004) and Mostafa (2005). It furnishes the two additional constraints to the problem:

$$T_{11} + T_{22} + T_{33} = 0 \tag{4.88}$$

$$T_{11}^2 + T_{22}^2 + T_{33}^2 = \frac{3}{2} \tag{4.89}$$

where T_{ii} with $i \in [1;3]$ represent the diagonal components of the reduced stress tensor.

However, the shape ratio does not appear explicitly in Eq. (4.87), yet it can be easily calculated from the angular parameter ψ. To note, this reduced stress tensor is expressed in the physical coordinate system, whereas the two previous ones are formulated in the systems of their eigenvectors. The present expression appears to be more complex and less intuitive than the two previous ones; it presents nevertheless the remarkable advantage of leading to a set of linear equations, which can be analytically solved in short computation times (Angelier and Goguel, 1979).

This sort of reduced stress tensor forms the backbone of the so-called *direct inversion methods* (e.g. Angelier and Goguel, 1979; Angelier, 1979, 1990, 2002; Sperner et al., 1993; Mostafa, 2005). Direct inversion methods are particularly elegant and efficient methods and will be addressed in more detail in Section 4.2.3.

Methods based on the types of reduced stress tensors as the ones given in Eqs (4.85) and (4.86) are normally employed in conjunction with nonlinear equations, which in turn demand relatively heavy iterative schemes. The corresponding methods are referred to as *grid search methods* (e.g. Carey and Brunier, 1974; Angelier, 1975, 1984; Etchecopar et al., 1981; Gephart and Forsyth, 1984; Choi, 1991; Delvaux and Sperner, 2003; Yamaji, 2000; Sato and Yamaji, 2006a).

'Grid search' stands for exploring the 4D parameter space until finding the best-fitting reduced stress tensor. In practice, large numbers of reduced stress tensors are successively tested in varying systematically the four stress parameters. For each tested tensor, shear stress vectors are computed for all input faults and compared to their respective fault slips using some prescribed misfit criterion (see examples in next paragraph). The optimal tensor is found when the sum of misfits reaches a minimum value.

Exploring the whole 4D parameter space is, however, expensive in terms of computation time. To lighten the procedure, the grid search is often combined with stochastic (e.g. Armijo and Cisternas, 1978; Armijo et al., 1982) or Monte Carlo (e.g. Etchecopar et al., 1981) approaches, which consist in identifying most probable locations of the solution in the parameter space, by means of repeating random tries, and focussing the exploration on these specific regions.

c) Misfit criteria and minimisation functions

The goal of the inversion is to find the stress tensor that minimises misfits between predicted shear stress vectors and observed fault slips. To these aims, one needs to define a *misfit criterion* in order to quantify the degree of goodness of the result. When various data are involved in the computations, a function based on the selected misfit criterion is constructed to quantify the average goodness of the result. In fault slip inversion methods, the latter is in general a function whose minimum value indicates the optimal stress tensor, that is a function that the algorithm tries to minimise or, in other words, a *minimisation function*. The minimisation functions listed below were elaborated for grid search methods. Some methods consider, however, the opposite strategy and involve *maximisation functions* (e.g. Yamaji, 2003; Otsubo and Yamaji, 2006; Yamaji et al., 2006; Žalohar and Vrabec, 2007), that is, these methods aim at maximising fitness. A maximisation function is easily defined as complementary to one of a minimisation function (see examples p. 981 in Yamaji et al., 2006).

The most natural misfit criterion of the fault slip inversion problem is the unsigned angle α separating shear stress vector from fault slip (Fig. 4.9), and ranging from 0° (i.e.

perfect fit) to 180° (i.e. fully inconsistent result). The corresponding minimisation function (e.g. Etchecopar et al., 1981; Yamaji, 2000) is built as the sum of all the misfits:

$$F_0 = \sum_{i=1}^{N} \alpha_i \tag{4.90}$$

where N represents the total number of fault slip data.

Although the physical meaning of the latter minimisation function is straightforward, computation of α_i from Eqs (4.70), (4.81), or from equivalent formulas, requires computation of an inverse trigonometric function, for example:

$$\alpha_i = \sin^{-1}\left(\vec{\tau}_i . \vec{u}_i\right) \tag{4.91}$$

where as previously stated $\vec{u}_i$ and $\vec{\tau}_i$ are unit vectors, thus the computation requires prior normalisation of $\vec{\tau}_i$ in addition.

The incorporation of inverse trigonometric functions results in a set of nonlinear equations, which demand quite heavy grid search strategies. In order to avoid computationally demanding algorithms, which were difficult to handle by most computers at the time, early fault slip inversion methods relied on misfit criteria and minimisation functions built on trigonometric functions, as the one deriving directly from Eq. (4.17) and employed by Carey (1976, 1979):

$$F_1 = \sum_{i=1}^{N} \left(\vec{\tau}_i \cdot \vec{u}_i\right)^2 = \sum_{i=1}^{N} \cos^2 \alpha'_i \tag{4.92}$$

where α'_i is the complementary angle to α_i (Fig. 4.9).

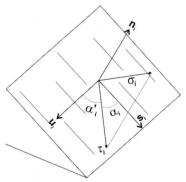

FIG. 4.9 Conventions used in text. α_i and α'_i represent angular separations between computed shear stress (i.e. $\vec{\tau}_i$) and observed fault slip (i.e. $\vec{s}_i$), and between computed shear stress and direction of zero slip (i.e. $\vec{u}_i$), respectively. As in Fig. 4.8, $\vec{\sigma}_i$, is stress vector and $\vec{n}_i$ fault plane normal.

The very first minimisation functions were, however, unable to discriminate slip sense. The drawback was quickly eliminated by the introduction of alternative minimisation functions, such as (Angelier, 1977):

$$F_2 = \sum_{i=1}^{N} \sin^2 \frac{\alpha_i}{2} \tag{4.93}$$

which may be re-written in the more pleasant form

$$F_2 = \frac{N}{2} - \frac{1}{2} \sum_{i=1}^{N} \cos \alpha_i \tag{4.94}$$

With the aim of optimising computation speed, other minimisation functions were later proposed, one may retain for example the minimisation function (Angelier, 1979; Angelier and Manoussis, 1980):

$$F_3 = \sum_{i=1}^{N} \tan^2 \left[\min \left(\alpha_i; \frac{\pi}{4} \right) \right] \tag{4.95}$$

for which calculations are drastically simplified for all misfit angles higher than 45°, that is, the angle above which misfits are too large to be accepted.

The minimisation function F_3 (and equivalent formulations, e.g. Yamaji et al., 2006) presents the additional advantage of suppressing the influence of data associated with large α_i angles, i.e. outliers, on the final result.

As they are formulated in the examples listed above, the minimisation functions offer the possibility of introducing easily ponderation schemes (e.g. Delvaux and Sperner, 2003). For example, in order to account for data quality, the minimisation function F_2 (Eq. (4.93)) may be re-written:

$$F_{2p} = \sum_{i=1}^{N} \omega_i \sin^2 \frac{\alpha_i}{2} \tag{4.96}$$

where ω_i are weighting factors, either input by the user or automatically attributed by the algorithm depending on certain quantitative criteria.

Weighting the data in function of quality may improve the results. However, it is not straightforward to select arbitrary numerical values for the ponderation factors and to attribute them to (semi)qualitative inferences (e.g. estimated measurement errors, estimated confidence on the inferred sense of fault slip, etc.). The reader should remain aware that data weighting is fundamentally a subjective operation.

Numerous minimisation functions were proposed following the pioneering works of Evelyne Carey and Jacques Angelier. An exhaustive catalogue of all of them would be somewhat superfluous, as all these functions rely on the same physical principle (i.e. the main Wallace–Bott hypothesis), albeit differ on the technical aspects of the adopted computation strategies (see an overview p. 211 in Célérier et al., 2012). For the sake of completeness, let us mention briefly two particular types of minimisation functions before closing the discussion.

Instead of minimising directly or indirectly α_i, some functions aim at minimising the rotation needed to bring the fault plane to the geometrical situation, where fault slip and computed shear stress agree (e.g. Armijo and Cisternas, 1978; Armijo et al., 1982; Angelier et al., 1982; Gephart and Forsyth, 1984; Gephart, 1990a, 1990b; Yin and Ranalli, 1993; Xu, 2004). The major advantage of this latter approach resides in the possibility of estimating the impact of all measurement uncertainties (i.e. on strike, dip and pitch) on the results.

The second particular type of minimisation functions involve vector formulations, which are equivalent to the simple angular criterion. This type of functions is mostly used in the formalism of direct inversion methods and will be detailed below in Section 4.2.3.

4.2.3 Direct inversion

a) Minimisation criterion

Although inversion of fault slip data is a problem relatively simple from the conceptual point of view, the equations to be solved become quickly rather lengthy and complex, as it is usual for problems in the 3D physical space. We saw that minimisation criteria based solely on the angular separation between shear stress and stria result in nonlinear equations that cannot be solved analytically. Their resolution demands heavy numerical strategies. Direct inversion was originally proposed as an alternative approach by Angelier and Goguel (1979) and Angelier (1979). The method was improved later by Angelier (1990) and Mostafa (2005). Direct inversion requires a set of linear equations leading to fast analytical resolution and, therefore, modification of the minimisation criterion and of the associated minimisation function.

Let us first examine the simple angular criterion, α_i, used in most minimisation functions, as discussed in the previous section, and depicted in Fig. 4.10A. This criterion

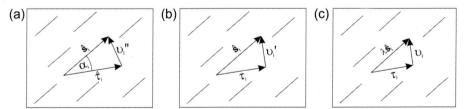

FIG. 4.10 (A) the classical angular separation minimisation criterion, α_i, and its equivalent vector formulation, $\vec{v}_i''$; (B) minimisation criterion taking into account calculated shear stress magnitude, the modulus of $\vec{v}_i'$ reaches its minimum for a given α_i when $\vec{v}_i' \perp \vec{\tau}_i$; (C) minimisation criterion following a similar rule as in (B), in the present case shear stress magnitude tends to the scale factor λ for small α_i values. $\hat{s}_i$ and $\vec{\tau}_i$ are fault slip and shear stress, respectively, circumflexes indicate unit vectors. After Angelier (1990).

ensures the minimum angular distance between shear stress and fault slip vectors but neglects shear stress magnitude, the two vectors being in general normalised in corresponding computation strategies. The approach presents two major drawbacks. First, the value of α_i becomes meaningless if τ_i is small, that is, if the stress vector, $\vec{\sigma}_i$, is nearly perpendicular to the fault plane. Subtle variations in the orientation of $\vec{\sigma}_i$ produce large variations in α_i, so that its value is in fact unreliable and the solution poorly constrained. Second, relatively low values of τ_i are physically inconsistent with fault reactivation, as most faults yield significant resistance to shear. In brief, the simple angular criterion is apparently incomplete.

The thoughts developed above conduct naturally to finding a new criterion that involves shear stress magnitude. Additionally, when considering shear stress magnitude, instead of the normalised shear stress vector, there is no more need for dividing shear stress components by polynomials (i.e. by the shear stress magnitudes themselves) in the set of Eq. (4.70). The equations to be solved become pleasantly linear.

Let us first remark that minimising α_i in the traditional inversion strategy is strictly equivalent to minimising the modulus of the vector $\vec{v}_i''$ (Fig. 4.10A) defined by:

$$\hat{s}_i = \hat{\tau}_i + \vec{v}_i'' \tag{4.97}$$

For instance, the minimisation function F_2 (Eq. (4.93)) may be re-written:

$$F_2 = \frac{1}{4} \sum_{i=1}^{N} v_i''^2 \tag{4.98}$$

We now consider shear stress magnitude (Fig. 4.10B), Eq. (4.97) is thus changed into:

$$\hat{s}_i = \vec{\tau}_i + \vec{v}_i' \tag{4.99}$$

The modulus of $\vec{v}_i'$ first decreases with τ_i until reaching a minimum value, when $\vec{v}_i' \perp \vec{\tau}_i$, and starts to increase if τ_i increases further. We remark that, for reasonably small values of α_i, the minimum value of v_i' indicates $\tau_i \approx s_i$ (that is, 1 in the present case). Thus, the criterion indicates an optimal solution if shear stress magnitude tends to some convenient value. This is particularly manifest if $\alpha_i = 0$, the ideal case $v_i' = 0$ corresponding to $\vec{\tau}_i = \hat{s}_i$. As a final remark, we should note that v_i' increases with α_i. In summary, the criterion conserves well the properties of the primitive angular criterion while adding a condition on shear stress magnitude.

The last step of the conception of a minimisation criterion adapted to direct inversion consists in scaling the vector relationship in a meaningful way. An astute choice is to attribute to the fault slip vector a modulus equal to the maximum shear stress magnitude, that is, half of the difference between the maximum and the minimum principal stress, named λ hereafter. The vector relationship is therefore (Fig. 4.10C):

$$\lambda \hat{s}_i = \vec{\tau}_i + \vec{v}_i \quad \text{with} \quad 0 \le v_i \le 2\lambda \tag{4.100}$$

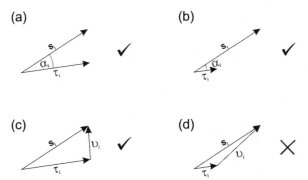

FIG. 4.11 Fundamental difference between the angular minimisation criterion, α_i, and the vector criterion, $\vec{v}_i$. α_i is kept constant for all depicted cases. (A) and (B) show acceptable agreement between fault slip and shear stress according to the angular criterion. (C) and (D) represent strictly similar situations as in (A) and (B), respectively, but evaluated with the $\vec{v}_i$ criterion. The case in (A) leads to a relatively modest modulus for $\vec{v}_i$ and is therefore acceptable. In turn, for the case in (B) the modulus of $\vec{v}_i$ is large indicating potential contradiction between shear stress magnitude and fault slip, i.e. shear stress is too low to produce fault slip. See text for details.

and its corresponding minimisation function is

$$F_4 = \sum_{i=1}^{N} v_i^2 \tag{4.101}$$

The new minimisation criterion, $\vec{v}_i$, presents obvious advantages compared to the angular criterion (Fig. 4.11). First, the value taken by α_i becomes meaningless if τ_i is small as discussed above, whereas v_i remains stable and close to λ. In brief, $\vec{v}_i$ is insensitive to tiny fluctuations in $\vec{\sigma}_i$ orientation. Second, although no explicit mechanical law is involved in the criterion, it meets the reasonable requirement of having sufficiently large shear stress magnitudes in order to overcome fault friction and, ultimately, in order to explain fault slip.

b) Solving the analytical problem

The choice of the minimisation criterion leads to linear equations but the finding of an analytical solution is not always straightforward and depends on the form selected for the reduced stress tensor. The reduced stress tensor adopted by Angelier (1990) is the one discussed previously (i.e. Eq. (4.87)), whose diagonal values are related through Eqs (4.88) and (4.89).

In the following, we will avoid lengthy analytical developments, which are superfluous to understand the mathematical foundations of the method. The interested reader may find complementary information on the mathematics in appendix of Angelier (1990). Let us first express the quantity v_i^2 used in the minimisation function, F_4. According to Eq. (4.100):

$$v_i^2 = \lambda^2 + \tau_i^2 - 2\lambda\tau_i \cos\alpha_i \tag{4.102}$$

which for the sake of the discussion is alternatively written

$$v_i^2 = \lambda^2 + \tau_i^2 - 2\lambda \hat{s}_i \cdot \vec{\sigma}_i \qquad (4.103)$$

Examining the system of Eq. (4.69), we conclude that the second term on the right side of Eq. (4.103), i.e. τ_i^2, is linearly dependent on the components of the stress tensor, the coefficients of the linear relationship being polynomials of the director cosines of the normal of the fault plane. Similarly, a simple glance at Eq. (4.65) convinces us that the term $\hat{s}_i \cdot \vec{\sigma}_i$ is also linearly dependent on the stress tensor components, the coefficients being polynomials of the components of the normal and of the fault slip vector, in the present case. Remember that using the stress tensor given by Eq. (4.87), the number of unknowns indicated in the general Eqs (4.65) and (4.69) is reduced to four.

In conclusion, v_i^2 can be written as a linear expression of the components of the stress tensor (and therefore of the four unknowns of the problem), involving simple polynomials built on the fault slip datum gathered in the field (i.e. fault attitude and stria orientation). By extension, the sum of all v_i^2, that is the minimisation function F_4, can also be written as a linear function of the components of the stress tensor, yet the coefficients of this function take relatively complex forms of sums of polynomials. Extreme values of F_4 are searched for by calculating its partial derivatives with respect to the four unknowns of the stress tensor and annulling them:

$$\frac{\partial F_4}{\partial d} = 0 \quad \frac{\partial F_4}{\partial e} = 0 \quad \frac{\partial F_4}{\partial f} = 0 \quad \frac{\partial F_4}{\partial \psi} = 0 \qquad (4.104)$$

After development, the four partial derivatives indicated above result in the system of linear equations:

$$
\begin{aligned}
&Ad + De + Ef = G\cos\psi + J\sin\psi + U \\
&Dd + Be + Ff = H\cos\psi + K\sin\psi + V \\
&Ed + Fe + Cf = I\cos\psi + L\sin\psi + W \\
&M\cos 2\psi + \frac{1}{2}N\sin 2\psi = (Gd + He + If + Q)\sin\psi - (Jd + Ke + Lf + P)\cos\psi
\end{aligned}
\qquad (4.105)
$$

where A, B, C... U, V, W are sums of polynomials constructed on the N field data and ψ, e, d and f the four unknowns of the reduced stress tensor as shown in Eq. (4.87).

The system of Eqs (4.105) is analytically solved and in most cases yields the two extreme values of F_4, corresponding themselves to two (ψ, e, d, f) sets. The set of values associated with the minimum of F_4 is, evidently, solution of the problem and characterises the reduced stress tensor that best explains the fault slip data.

It is necessary to attribute a numerical value to λ prior computations. As stated previously, λ is defined as the maximum shear stress, whose value depends on the components of the adopted stress tensor. The stress tensor given in Eq. (4.87) presents the advantage of simplifying the algebra when formalising the direct inversion problem. However, stress magnitude, thus maximum shear stress, is related to the orientations of the principal axes of stress for this particular form of tensor, so that λ cannot be fixed a priori.

In order to overcome the difficulty, Angelier (1990) proposed to adjust the value of λ through successive stress tensor determinations. For each tensor determination the maximum shear stress computed during the previous step is assigned to λ. The final step assumes the particular form:

$$T_R = \begin{pmatrix} \cos\psi & 0 & 0 \\ 0 & \cos\left(\psi+\dfrac{2\pi}{3}\right) & 0 \\ 0 & 0 & \cos\left(\psi+\dfrac{4\pi}{3}\right) \end{pmatrix} \tag{4.106}$$

for which $\lambda = \frac{\sqrt{3}}{2}$.

Similarly to the case of the angular minimisation criterion, where the quality of the fit is evaluated through an arbitrary choice of the maximum angular separation that is acceptable, a misfit threshold needs to be defined. To these aims, Angelier (1990) introduced the 'ratio upsilon', which is the ratio of v_i to the largest shear stress expressed in percentage:

$$\text{RUP} = 100 v_i \bigg/ \frac{\sqrt{3}}{2} \tag{4.107}$$

The RUP estimator varies from 0%, indicating maximum shear stress parallel to fault slip and with same sense (i.e. perfect fit), to 200%, corresponding to maximum shear stress parallel to fault slip but with opposite sense (i.e. largest misfit). The quality of the fit is good if $\text{RUP} \leq 50\%$, (potentially) acceptable if $50\% < \text{RUP} \leq 75\%$, and poor otherwise. An average RUP may be calculated for all the dataset. However, the average RUP is not informative enough about the quality of the stress tensor determination, and it is normally preferred to check the relative number of fault slip data falling in each category (see Angelier, 1990 for details).

c) Advantages and refinement of the direct inversion method

In terms of computation time, direct inversion is much faster than grid search, a benefit that nowadays is not as important as it was in the 20th century, when the method emerged. An enduring advantage of the method, which stems from its specific minimisation function, is its better accuracy with respect to traditional fault slip inversion strategies. Let us consider the set of fault slip data depicted in Fig. 4.12 to demonstrate the latter statement. Minimisation of the shear-slip angle results in ambiguous determination of the orientations of σ_1 and σ_3. Indeed, any one of the configurations for the stress axes shown in Fig. 4.12A honours equally well the minimisation problem, that is, the minimisation function e.g. F_2 (i.e. Equation (4.93)) equals zero for all depicted stress configurations. However, employing the minimisation criterion based on $\vec{v}$, the inversion of the fault slip data leads to a unique configuration for the stress axes, the only one that maximises shear stress for all faults (Fig. 4.12B).

The direct inversion method of Angelier (1990) was later adopted by Sperner et al. (1993), Angelier (2002), Mostafa (2005) and Sasvári and Baharev (2014). In particular, Mostafa (2005) improved the original method and demonstrated that the choice of an

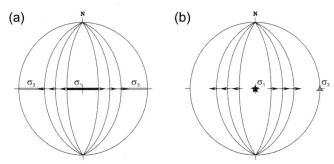

FIG. 4.12 Determination of the orientations of the principal axes of stress based on (A) the angular minimisation criterion and (B) the $\vec{v}$ minimisation criterion. In the first case, numerous configurations for the stress axes honour equally the problem, so that the orientations of σ_1 and σ_3 are not totally constrained (*black* and *grey* strips depict possible orientations for σ_1 and σ_3). Applying the $\vec{v}$ minimisation criterion, only one configuration for the stress axes leads to shear stress maximisation for all the faults, demonstrating that the latter criterion is more powerful than the angular one. The fault slip data are artificial, note that the shape ratio cannot be determined with this dataset (i.e. any Φ value is consistent with the dip slip movements for the present stress configuration, see discussion in Section 4.1.2b), independently of the choice made for the minimisation criterion. Projection is on the lower hemisphere of a Schmidt net.

arbitrary λ value was at the origin of minor inaccuracies. While testing the direct inversion method on sets of synthetic fault slip data, generated using the Bott equation (see e.g. Equation (4.13)), he discovered mismatches of few degrees between predicted shear stress vectors and fault slips (corresponding to nonzero RUP values of less than 50%). In forcing shear stresses to tend artificially to the maximum permissible value λ, a condition met only for fault planes bisecting the σ_1-σ_3 angle, the original method resulted in slight discrepancies. To remediate to the problem and to demonstrate its cause, Mostafa modified Angelier's approach with additional computation loops, in order to release the too stringent λ constraint. As first step, Mostafa's method determined an initial reduced stress tensor with the original direct inversion of Angelier (1990). In the following step, the tensor was used to forward calculate maximum shear stresses for all faults and the results were stored as collection of individual λ_i. A new inversion step was performed with the condition that shear stress on each fault approached its corresponding λ_i value, as computed during the forward computation step. The two last steps were reiterated until the λ_i values and the four parameters of the reduced stress tensor stabilised. The tests on synthetic data carried out by Mostafa (2005) showed that his modified direct inversion was able to determine the originally prescribed stress tensor perfectly, with individual RUP values lower than 1% and equal to 0% for most of them.

4.2.4 Incorporation of friction

a) Friction and mechanical compatibility of fault slip data

As discussed in Section 4.2.3, a small angular separation between shear stress and fault slip does not ensure that the fault is mechanically compatible with the computed stress tensor. Fault slip is only possible if shear stress magnitude is high enough to over-come the resistance to shear of the discontinuity (see Section 3.2.2). Fault slip data pre-senting acceptable angular misfits with predicted shear stress vectors but being mechanically incompatible with the computed stress tensor (taking into account reason-able friction criteria) are seemingly products of other stress conditions, i.e. either related to other tectonic phases or reflecting local stress alterations (e.g. Fig. 7.10). This potential drawback was identified early in the development of fault slip inversion methods, and the very first approaches integrating friction in the analysis (e.g. Etchecopar, 1984; Gephart and Forsyth, 1984; Sassi and Carey-Gailhardis, 1987; Gephart, 1990a; Fleischmann and Nemcok, 1991) consisted in checking the mechanical consistency of the fault slip data a posteriori and with the help of Mohr diagrams (Fig. 4.13).

The mathematical analysis of Célérier (1988) added further support to the importance of involving friction in fault slip inversion. Célérier (1988) extended the theoretical study of McKenzie (1969) on the relationship between nodal planes derived from earthquake focal mechanisms and orientations of the principal axes of stress (see definitions and details in Sections 4.3.1 and 4.3.2). He concluded that, in general, the orientations of the principal axes of stress were better resolved, even from a single datum, when friction was consid-ered in the fault slip inversion scheme.

The methods that evolved from the pioneer studies of Etchecopar (1984) and Gephart and Forsyth (1984) attempted to fully integrate fault friction in their inversion algorithms. These may be classified according to two main objectives: (1) methods aiming at improv-ing the determination of the parameters of the stress tensor (e.g. Reches, 1987; Hardcastle, 1989; Hardcastle and Hills, 1991; Phan-Trong, 1993; Žalohar and Vrabec, 2007) and (2) methods searching for quantifying friction (Reches et al., 1992; Yin and Ranalli, 1995).

Because the friction condition is expressed through an inequality (i.e. fault slip along a preexisting discontinuity occurs when shear stress is equal or overcomes the term to the right in Eq. (3.110)), the problem is nonlinear and iterative grid search methods were employed to solve it. An exception is the semianalytical approach of Reches (1987) and Reches et al. (1992), where the condition is arbitrarily reduced to strict equality. To note, most methods simplify Eq. (3.110) by ignoring cohesion. The assumption appears reason-able, cohesion being negligible for stress magnitudes typical of the upper crust.

Introducing friction in traditional fault slip inversion algorithms has a similar effect on the results as employing direct inversion methods (Section 4.2.3). That is, shear stress magnitudes are required to be large enough to explain faulting. The only difference between both approaches resides in the strict threshold (i.e. the prescribed shear resis-tance, which in most cases is reduced to the angle of friction) that the ratio shear to normal stress has to overcome for the fault slip datum to be acceptable. The advantage of incor-porating friction resides mainly in the capacity of the method to eliminate spurious data, which might bias significantly the determination of the tensor otherwise. Fortunately, the mismatch between results obtained with classical fault slip inversion methods and the ones delivered by methods involving friction tends to be negligible, when sufficiently large numbers of data with acceptable variety are inverted (e.g. Fig. 4.13).

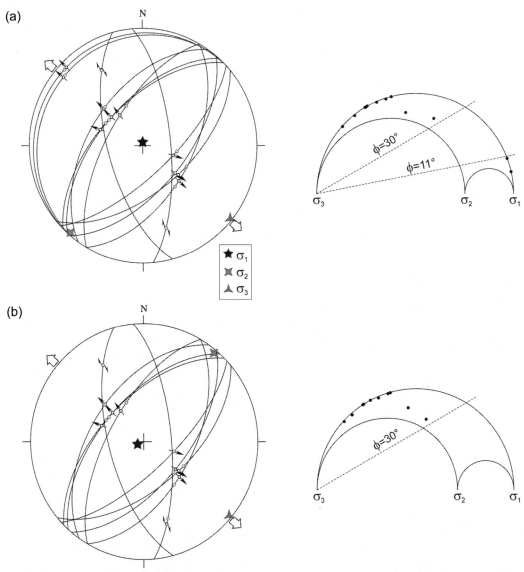

FIG. 4.13 Results of fault slip inversion of synthetic data represented on lower hemispheres of Schmidt nets and Mohr diagrams. The inversions were performed with the grid search method Brute (Sasvári and Baharev, 2014), which involves the angular separation between input fault slip and computed shear stress as misfit criterion (acceptable individual misfits up to 30° in the present case). (A) Solution calculated with the whole dataset, including two nearly horizontal faults depicted in *red*. The misfit criterion indicates an excellent agreement between the two red faults and the computed stress tensor, as suggested by the negligible separations between the poles of the striae and their corresponding shear stress vectors (depicted as yellow poles, senses are concordant with the fault slips but not indicated for the sake of clearness). The representation of the results in the Mohr space suggests that most of the fault slip data are compatible with reasonable friction criteria (i.e. most poles plot well above the friction line $\phi = 30°$) save the two in red, whose poles fall below the line $\phi = 11°$, i.e. the corresponding fault planes should possess unreasonably low friction values to be reactivated in the computed stress state. (B) For comparison, results of the computation after excluding the two nearly horizontal faults. Note that in the present case the obtained stress tensor is nevertheless very similar to the one calculated in (A).

Admittedly, involving friction in the computation scheme is more satisfactory from the physical point of view. However, it is done at the expense of introducing an additional assumption, that is, an arbitrary value for the angle of friction. Most authors select standard values of ~30° as suggested by rock mechanics experiments. A group of authors (Reches, 1987; Reches et al., 1992) explore different angles of friction and choose the optimal result on rather subjective grounds.

b) Constraining a fifth parameter of the stress tensor

An interesting outcome of adding a mechanical condition to the fault inversion problem is, nevertheless, the possibility of constraining a fifth parameter of the reduced stress tensor (e.g. Bergerat et al., 1982, 1985; Angelier, 1983; Sassi and Carey-Gailhardis, 1987; Célérier, 1988). In substance, this fifth parameter quantifies the relative roles of differential stress and pressure in triggering fault slip, the former promoting fault slip whereas the later impeding it (see also Section 8.2).

The interplay between the two quantities is particularly well imaged in the Mohr space, either in the case of failure, where the failure line has to be tangent to the Mohr circle (Fig. 4.14A), or in the case of reactivation, where a portion of the Mohr circle should stand above the friction line in general (Fig. 4.14B). Either condition is honoured by an infinity of Mohr circles provided they all depict the same geometrical relationship with respect to the line representing the shear resistance of the material (i.e. either failure or friction line). In other words, finding a Mohr circle solution of the problem is equivalent to adjusting both its diameter (i.e. differential stress) and its position (i.e. which in turn is equivalent to adding or subtracting a pressure term), in order that the circle respects the geometrical constraint. The reasoning leads us to conclude that the fifth parameter corresponds to some ratio of a term representing directly or indirectly differential stress (that is, the diameter of the Mohr circle) to a term representing directly or indirectly a pressure (that is, the position of the Mohr circle), or alternatively the inverse ratio.

Examining Fig. 4.14C, where two Mohr circles honouring the failure criterion are depicted, it is straightforward to demonstrate that triangles *abc*, *bcd* and *abd* are homothetic to triangles *aij*, *ijk* and *aik*, respectively. Thus, it is possible to derive various proportionality ratios that remain constant from one circle to the other, for example:

$$S = \frac{bc}{ad} = \frac{ij}{ak} = \frac{1}{2} \frac{\sigma_1 - \sigma_3}{\sigma_1 + \dfrac{C_i}{\tan \varphi_i}} \tag{4.108}$$

or

$$S_G = \frac{bc}{ac} = \frac{ij}{aj} = \frac{\sigma_1 - \sigma_3}{\sigma_1 + \sigma_3 + \dfrac{2C_i}{\tan \phi_i}} \tag{4.109}$$

where C_i and ϕ_i are cohesion of intact rock and angle of internal friction, respectively (see Section 3.2.2).

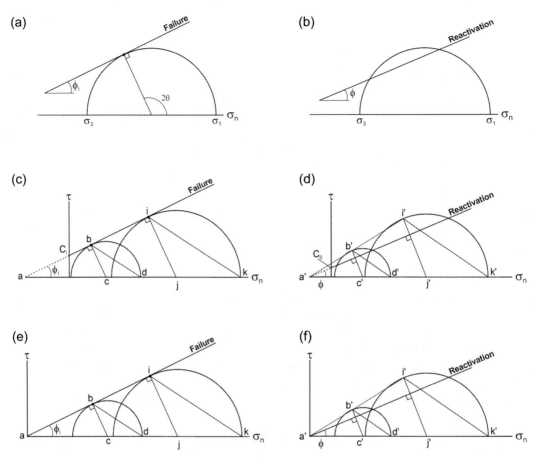

FIG. 4.14 Determination of a fifth parameter of the reduced stress tensor taking account failure/reactivation criteria. The left and right panels depict geometrical relationships between the Mohr circle and the failure and reactivation (i.e. friction) lines, respectively. (A) and (B) show the geometrical configurations than honour the mechanical criteria. In both cases only the ratio between scale and position of the Mohr circle (i.e. the fifth parameter) is constrained but not its precise location and radius. (D) and (C) depict homothetic relationships from which the fifth parameter can be derived (see text for details). (E) and (F) demonstrate that these relationships remain when the origin coincides with the point a (alternatively a'). C_i and ϕ_i are, respectively, cohesion of intact rock and angle of internal friction, whereas C_0 and ϕ are, respectively, cohesion and angle of friction of preexisting discontinuities (see Sections 3.2.2 and 3.2.3 for further details).

Both equations involve a measure of the scale of the Mohr circle divided by a measure of its relative position and, therefore, they are both expressions of the fifth parameter of the reduced stress tensor that the friction condition permits to solve.

However, the terms reflecting position depend on both cohesion and coefficient of friction, hence on two independent parameters. The analysis may be simplified considering that shifting the origin of the Mohr space to point a does not alter the previously evidenced geometrical relationships (compare Fig. 4.14C with Fig. 4.14E). The operation is equivalent to subtracting an isotropic component of magnitude $C_i/\tan\phi_i$ to the reduced stress tensor (Célérier, 1988). Hence, the operation affects neither the four parameters traditionally derived by means of fault slip inversion nor the fifth parameter.

In turn, Eq. (4.108) simplifies into:

$$S = \frac{1}{2} \frac{\sigma_1 - \sigma_3}{\sigma_1} \tag{4.110}$$

or, omitting the factor ½,

$$S_C = \frac{\sigma_1 - \sigma_3}{\sigma_1} \tag{4.111}$$

which is the expression of the fifth parameter adopted by Célérier (1988).

Eq. (4.109) becomes:

$$S_G = \frac{\sigma_1 - \sigma_3}{\sigma_1 + \sigma_3} \tag{4.112}$$

which is the ratio proposed by Gephart (1992) and specifically the half of the ratio of differential stress to (2D) mean stress. Alternative expressions of the fifth parameter may be found in e.g. Reches (1987), Angelier (1989), Yin and Ranalli (1995) and Choi et al. (1996), and a nonexhaustive list of them in Célérier et al. (2012).

As shown in Fig. 4.14D and E, the homothetic relationships pertaining to the failure criterion can be generalised to the reactivation one. Thus, the expressions of the fifth parameter are directly applicable to the case of fault reactivation, which is precisely the case addressed by fault slip inversion methods.

Finally, some approaches consider friction together with further constraints in order to reconstruct the six parameters of the stress tensor (Bergerat et al., 1982, 1985; Angelier, 1983, 1989; Choi et al., 2013). These are addressed in detail in Chapter 8 (see Section 8.2).

4.2.5 Incorporation of discontinuities without shear

a) Nonreactivated discontinuities

The idea of incorporating nonreactivated discontinuities in fault slip inversion (Angelier, 1992a) stems, obviously, from inversion of calcite twins, where not only twinned e-planes but also untwinned ones are crucial to constrain the solution (see Chapter 6). In virtue of the main Wallace–Bott hypothesis, fault slip inversion methods rely primarily on discontinuities reactivated in shear, and are easily extended to include the case of neoformed faults, which once formed are considered as 'preexisting' their last slips. Preexisting discontinuities that do not slip yield, however, usable information to the inversion as well. Assuming standard frictional properties, their lack of response to the imposed stress is suggestive of relatively low shear stress to normal stress ratios.

As discussed in the previous section, accounting for friction renders fault slip inversion more satisfactory from the mechanical point of view, but at the expense of heavy iteration schemes and additional assumptions. In contrast, the approach adopted by Angelier (1992a) simplifies the friction constraint, as given by nonreactivated discontinuities, to

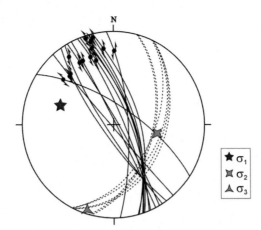

FIG. 4.15 Example of paleostress reconstruction from 14 fault slip data supplemented with 6 nonreactivated discontinuities (i.e. joints, depicted here with *dashes*), Φ ∼0.5. Note that the fault slip data show rather poor variety. Original data and results are from Angelier (1992a). Projection on the lower hemisphere of a Schmidt net.

the condition that their associated shear stresses reach magnitudes as low as possible. In brief, the method he proposed, an extension of his direct inversion method (see Section 4.2.3), searches for maximising shear on reactivated and/or neoformed discontinuities but attempts to minimise shear on nonreactivated ones.

In the case of favourable fault slip datasets, introducing nonreactivated discontinuities does not improve the final result (see figure 1 in Angelier, 1992a). The approach is however advantageous when the fault slip dataset at hand shows limited variety (Fig. 4.15) and leads to a poorly constrained solution. For example, the 14 NW-SE sinistral strike-slip faults measured in Taiwan (Angelier, 1992a) and depicted in Fig. 4.15 are rather similar, that is, each datum furnishes roughly the same information and the whole dataset constrains badly the tress tensor. Supplementing the fault slip data with 6 preexisting joints, which were not reactivated, permits to better determine the stress tensor. Noteworthy, the solution corresponds to an average RUP estimator of 25% (see Section 4.2.3), indicating a general good fit between fault slips and computed shear stresses.

b) Mode 1 fractures and stylolithes

Mode 1 (i.e. extension) fractures and stylolithes (i.e. compression structures) furnish information on the stress under which they originated (see Sections 2.2.2 and 2.2.4, respectively). The former open perpendicular to the minimum principal stress, whereas the latter form orthogonal to the maximum principal stress, in most cases (for more details see Section 5.1 and Section 8.4). It is thus tempting to couple fault slip inversion with inversion of these nonshear discontinuities in order to refine the determination of the reduced stress tensor.

The most obvious approach consists in finding a tensor whose one of the two extreme principal axes of stress (i.e. either σ_1 or σ_3, depending on the type of discontinuity) is parallel to the normal of the discontinuity. Corresponding minimisation functions are, for example, the ones searching for minimising the difference between the stress vector computed for the discontinuity and the extreme principal axis of stress relevant to the structure (Angelier, 1992b), that is:

$$\text{For mode 1 fractures} \quad F_T = \sum_{i=1}^{N} \left(\vec{\sigma} - \vec{\sigma}_3 \right)^2 \tag{4.113}$$

$$\text{For stylolithes} \quad F_P = \sum_{i=1}^{N} \left(\vec{\sigma} - \vec{\sigma}_1 \right)^2 \tag{4.114}$$

where N is number of data.

Both functions minimise shear stress while forcing the modulus of the stress vector to tend to either σ_3 or σ_1. These minimisation functions are easy to implement in fault slip inversion programs but require iterative algorithms. Similar approaches, albeit with slightly different minimisation functions, were later adopted by Delvaux and Sperner (2003), in their fault slip inversion code, and by Maerten et al. (2016b), in their geomechanical paleostress inversion method (see Section 7.3 and specifically Eqs (7.44) and (7.45)).

As it is the case for nonreactivated discontinuities, incorporation of mode 1 fractures or stylolithes in the inversion does not improve significantly the final result, if enough fault slip data with satisfactory variety are available. Adding such structures represents an advantage when the tensor cannot be reasonably constrained from the fault slip data alone.

For instance, Fig. 4.16A depicts a rather typical case of a nearly uniform fault slip dataset (measured in Taiwan by Angelier, 1992b). Moreover, the sample contains only four data. The lack of conjugate faults and/or distinct fault orientations implies that the reduced tensor obtained after inversion is poorly constrained. The addition of mode 1 fracture data, measured at the same site than the strike-slip faults and presumably coeval with them, brings strong constraints on the orientations of the principal axes of stress (Fig. 4.16B). The average RUP estimator is 24% (see Section 4.2.3), representative of a good result. In particular, the improvement of the fit is suggested by the final location of the poles of the measured faults in the Mohr space, the poles adopting positions consistent with neoformed ruptures in standard friction conditions. Note however that the shape ratio remains poorly determined in the present case (i.e. σ_2 is parallel or nearly parallel to the strike-slip faults and, expectedly, to the extension fractures).

4.2.6 Separation of heterogeneous fault slip data

a) The problem of polyphase deformation

Commonly, rocks experience successive deformation phases through time, and one expects that the older are the rocks the more complex is the accumulated deformation. Therefore, fault slip data collected in the field involve often subsets of data associated with

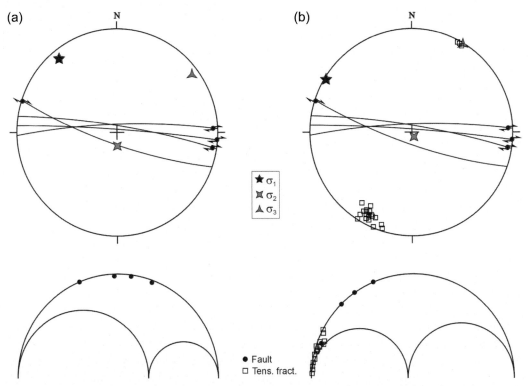

FIG. 4.16 (A) paleostress determination from fault slip data; in the absence of conjugate faults and/or more variety in the dataset, the orientations of the principal axes of stress and the shape ratio are poorly constrained. (B) paleostress determination with the same fault slip data supplemented with coeval mode 1 fracture data. The solution is significantly improved as suggested by e.g. the position of the fault poles in the Mohr diagram. Original data and results are from Angelier (1992b). The values of the shape ratio, Φ, are suggested by the Mohr diagrams. Projection on the lower hemisphere of a Schmidt net.

distinct deformation phases or paleostress states. Such data collections are referred to as *heterogeneous* (or *polyphase*) *fault slip data* in opposition with *homogenous* (or *monophase*) *fault slip data*, the latter corresponding to a unique paleostress tensor.

Inversion of a heterogeneous dataset leads evidently to a meaningless 'bulk' paleostress tensor. Instead, a sensible strategy consists in splitting the heterogeneous dataset into homogenous subsets in order to compute a single stress tensor for each of them, that is, to identify the different stress events. Note that a stress event may not represent a *tectonic phase* (i.e. a regional-scale uniform stress imposed during a significant period of time); only the repetition of comparable stress states at different sites and additional geological documentation permits to relate it to tectonic processes.

Separation of heterogeneous fault slip data (and structural data in general) can be done manually and without notable difficulty by the experienced geologist. The remaining outliers in the subset are simply identified as the ones yielding unacceptable misfits after inversion, and a final inversion is carried out once the outliers have been removed and potentially incorporated in other subsets (Carey, 1979).

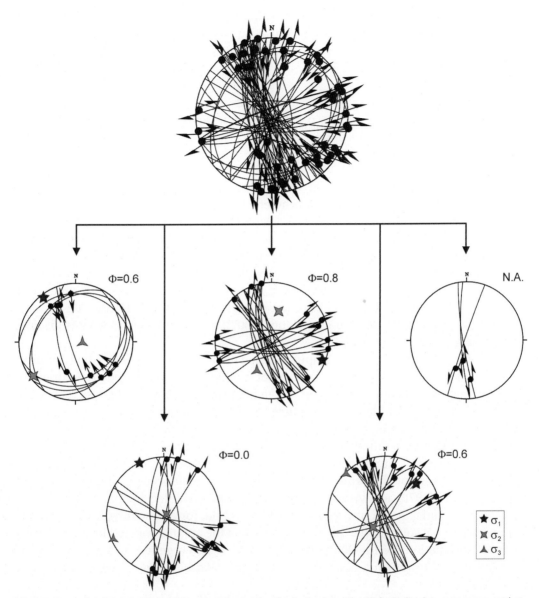

FIG. 4.17 Example of automatic sorting of heterogeneous fault slip data. The 80 fault slip data were measured at Ogmore, Wales, UK, and processed by Lisle and Vandycke (1996) using the method of Nemcok and Lisle (1995), for the separation, and the direct inversion method of Angelier (1990) for paleostress computations. Four stress events were identified in the analysis. Note the negligible number of nonattributed (N.A.) fault slip data.

It is, however, tempting to transfer these rather simple but sometimes rebarbative tasks to computers, as shown in the example depicted in Fig. 4.17. Computer-based separation algorithms exist since the infancy of fault slip inversion methods (Angelier and Manoussis, 1980; Etchecopar et al., 1981; Angelier, 1984). These may be classified in two main groups: methods coupling data separation and paleostress computations (e.g. Angelier and

Manoussis, 1980; Etchecopar et al., 1981; Galindo-Zaldivar and González-Lodeiro, 1988; Yamaji, 2000, 2003; Shan et al., 2003; Yamaji et al., 2006; Tranos, 2015, 2017), and methods separating the data before computation of the paleostress tensors (e.g. Carey-Gailhardis and Mercier, 1987; Nemcok and Lisle, 1995; Nemcok et al., 1999; Delvaux and Sperner, 2003).

Numerous methods to separate heterogeneous data have been developed (see review p. 213–214 in Célérier et al., 2012), presenting all of them is certainly redundant. In the following, we restrict the discussion to some classical examples belonging to each of the categories listed above.

b) Data separation coupled to paleostress tensor computations

Various methods exist to separate heterogeneous fault slip data while computing paleostress tensors. Although these methods involve different mathematical and numerical strategies, they all rely primarily on the angular separation between fault slip and shear stress (e.g. Fig. 4.9) to sort out the data. That is, the objective of the computations is to associate reduced stress tensors with subsets of fault slip data, such as angular misfits are minimum. To note, few methods require, in addition, the data subsets to remain compatible with a prescribed friction criterion (e.g. Hardcastle, 1989; Hardcastle and Hills, 1991; Phan-Trong, 1993; Žalohar and Vrabec, 2007), in a similar way as what is discussed in Section 4.2.4.

One of the simplest forms of computer-based data separation was originally proposed by Carey (1979). In this approach, subsets of homogenous fault slip data are identified with the help of histograms depicting the angular misfits after initial stress tensor computation (see figure 5 in Carey, 1979). Once identified the homogenous subsets are manually separated and their corresponding paleostresses calculated.

More automated methods take advantage of their specific grid search algorithms, which are based on numerous trial-and-error attempts and subsequent exploration of the parameters of the reduced stress tensor. For instance, the method of Armijo and Cisternas (1978) and Armijo et al. (1982), later adapted by Phan-Trong (1993), rotates systematically the principal axes of stress and, for each configuration, searches for an acceptable solution (i.e. the one minimising the angular misfits) by means of considering fluctuations of the shape ratio. In case of heterogeneous fault slip data, the method is able to return a collection of tensors, each one explaining a subset of the input data.

In a similar manner, the methods of e.g. Etchecopar et al. (1981), Etchecopar (1984), Galindo-Zaldivar and González-Lodeiro (1988), Hardcastle (1989) and Hardcastle and Hills (1991) explore large numbers of tensors to detect the ones that fit the best subsets of data, whose initial critical size is prescribed. The exploration is carried out systematically on regularly spaced grids save for Etchecopar's methodologies, where the tested stress tensors are selected randomly in order to reduce computation time. Rapid increase in computing power in the 2000s allowed for extending these previous methodologies to very fine search grids. For example, Yamaji (2003), Yamaji et al. (2006) and Žalohar and

Vrabec (2007) map systematically maximisation functions and detect peaks to identify homogeneous subsets.

Another classical approach to separate heterogeneous fault slip data is clustering. *Clustering* consists in grouping objects according to similarity or, alternatively, in separating objects that are dissimilar. Thus, a group of similar objects is a *cluster*. In the present case, the 'objects' are the fault slip data and 'similarity' refers to the value of the misfit angle. Accordingly, the method of Carey (1979) corresponds to semimanual clustering. Automatic computer-based clustering is an iterative process and therefore requires grid search algorithms.

The first automatic clustering method was developed by Angelier and Manoussis (1980) and later modified by Huang (1988) and Huang and Angelier (1989). The method is an adaptation of the dynamic clustering technique of Diday (1971), which consists in aggregating objects around nuclei while searching for minimising variances inside each aggregate and refining the nuclei iteratively. The main difference with Diday's technique resides in the need of computing stress tensors and, specifically, misfit angles that serve as measures of variance, at each iteration of the clustering. The steps of the algorithm of Angelier and Manoussis (1980) are the following:

1. define initial nuclei, preferentially each nucleus should contain fault slip data suspected to be related to a common stress event;
2. calculate a reduced stress tensor from each data nucleus (inverse step);
3. compute misfit angles for each datum using successively each calculated tensor (forward step);
4. identify the minimum misfit angle for each datum to sort the data according to the stress tensor associated to the minimum misfits, and form new nuclei;
5. calculate new reduced stress tensors from the data nuclei defined in step 4 (inverse step);
6. repeat step 3 with the tensors calculated in step 5;
7. repeat step 4 with the results obtained in step 6, new nuclei are defined;
8. compare the nuclei obtained in step 7 with the ones defined in step 1;
9. if the nuclei are identical end the process, the data are separated in homogeneous subsets, if not restart from step 2 until convergence.

Simplicity and short computation times are the main advantages of the method of Angelier and Manoussis (1980). Heavier and, to some extent, more refined clustering techniques could be implemented on personal computers only two decades later. For example, the multiple inverse method of Yamaji (2000) is akin to clustering. This latter method searches for partitioning the heterogeneous fault slip dataset in a prescribed number of clusters of homogeneous data. Similarly to Angelier and Manoussis' method, the data are correlated to each other based on the tested reduced stress tensor and the resulting values for the misfit angles. However, Yamaji's method employs a generalised form of the *Hough transform*, which is a powerful mathematical tool originally developed for automatic detection of geometrical features in digital images (see Appendix A in Yamaji et al., 2006). Otsubo and Yamaji (2006) improved the method by inserting a preprocessing phase. As in Nemcok and Lisle (1995), fault slip data attributes are used to identify preliminary clusters during this preprocessing phase (see details in next paragraph). The improvement helps to eliminate spurious solutions. Finally, Otsubo et al. (2006) introduced automatic detection of the clusters using k-means clustering. In particular, this latter modification presents the advantage of permitting quantification of confidence levels.

To complete the nonexhaustive list of separation methods, one should mention graphical methods. The most ancient one is the astute y-R method of Simón-Gómez (1986). In short, 2D curves depicting shape ratio (i.e. R) against azimuth of the maximum principal stress in the horizontal plane (i.e. y) are drawn for each analysed fault. Crossing 'points' (i.e. taking into account reasonable uncertainty margins) between the curves identify homogenous subsets of data and their corresponding reduced stresses. Fry (1992b) proposed a spherical plot version of the y-R method, whereas Célérier and Séranne (2001) adopted a different approach in modifying the Breddin's graph method. The previous graphical methods are, however, limited to Andersonian stress states.

Graphical representation of the inverse problem in the 6D stress tensor parameter space allows for overcoming the previous limitation (Fry, 1999, 2001) and represents the background of the automatic separation methods of Shan et al. (2003, 2004), Shan and Fry (2005) and Yamaji et al. (2006). The representation of the inverse problem in the so-called Fry's σ-space will be addressed in more detail in Section 4.2.7.

c) Data separation prior paleostress tensor computations

The second class of methods involves automatic sorting of the fault slip data into homogenous subsets before computation of the reduced stress tensors. A straightforward and simple approach consists in testing data compatibility using the right dihedra method (e.g. Carey-Gailhardis and Mercier, 1987; Delvaux and Sperner, 2003). The right dihedra method was developed to estimate graphically the orientations of the principal axes of stress (see full details in Section 4.3). Thus, it may be used to discriminate fault slip data according to the possible stress orientations suggested by the analysis.

A more elaborated method is the hierarchical clustering of Nemcok and Lisle (1995). The hierarchical clustering of fault slip data is based on the degree of similarity between each pair of data, itself calculated on the basis of attributes attached to each datum. An example of application of the method is given in Fig. 4.17. The algorithm of Nemcok and Lisle (1995) proceeds as it follows:

1. define attributes for each fault slip datum; these attributes are coded as binary strings defining the degree of compatibility between each datum and each one of the tested reduced stress tensors;
2. compute similarity between each pair of fault slip data; briefly, similarity is quantified for each pair based on the number of common tensors explaining each datum of the pair and the number of tensors that do not fit any of the data of the pair (see equation (1) in Nemcok and Lisle, 1995);

3. start to aggregate pairs of data according to highest similarity;
4. continue the clustering process by decreasing similarity;
5. the clustering ends, evidently, in one cluster; however, eye inspection of the dendrogram of the analysis allows for identifying the most likely homogenous subsets.

In detail, compatibility between one datum and a stress tensor is judged upon the value adopted by the angular misfit between fault slip and computed shear stress. That is, the first step of the clustering involves forward calculation of series of shear stress vectors for each fault plane, considering high numbers of uniformly distributed reduced stress tensors. To note, Nemcok et al. (1999) replaced the initial testing of multiple stress tensors by a right dihedra analysis and expanded the method to treat calcite twins.

d) Concluding remarks

Liesa and Lisle (2004) tested the computer-based methods of Etchecopar et al. (1981), Nemcok and Lisle (1995) and Yamaji (2000) on artificial and natural heterogeneous datasets. They found that each method had its specific drawbacks, yet the results of the three methods showed broad agreement.

Yamaji et al. (2006) demonstrated that traditional angular separation criteria led almost systematically to spurious results, when used to separate relatively large sets of heterogeneous data. Their analysis showed that automatic separation of large numbers of artificial data succeeded in reconstructing predefined data clusters but, also, predicted spurious homogenous subsets, accidently constructed on relatively modest numbers of data.

In conclusion, methods to sort out heterogeneous data may be useful, especially when the database under scope is large. They may however generate more or less serious artefacts. Such methods need to be used with much caution, and their results have to be confronted to field data and carefully reviewed by the geologist, as recommended by numerous previous authors (e.g. Angelier, 1984; Delvaux and Sperner, 2003; Sperner and Zweigel, 2010; Célérier et al., 2012).

4.2.7 Geometrical representation of the inversion problem: The Fry's σ-space

a) Original definition of the σ-space (Fry, 1999, 2001)

We have seen in Section 4.2.1 that the fault slip inversion problem may be formalised as a geometrical problem in a 6D space. Carey (1976) demonstrated the mathematical validity of fault slip inversion in the space of the six components of the stress tensor (see in particular Section 4.2.1.e). Fry (1999, 2001) embarked on a similar analysis but with the specific aim of constructing graphical representations to visualise data and solutions easily. Expressing the fault slip inversion problem in what is commonly referred to as the *Fry's σ-space* (i.e. the space of the six components of the stress tensor) is equivalent to apply a mathematical transformation, which renders the problem linear and, thus, easier to manipulate than its nonlinear formulation in the physical space (see Section 4.2.2.a). As such, Fry's inversion scheme may be viewed of a specific type of direct inversion. Fry's elegant formulation of the inversion problem met much success during the two last decades and constitutes the background of the works of e.g. Shan et al. (2003, 2004, 2006), Shan and Fry (2005), Sato and Yamaji (2006a, 2006b) and Hansen (2013).

Although the equations of Fry are reminiscent of the ones presented in Section 4.2.1, we will follow his own notations for easier comparison with the original papers (Fry, 1999, 2001). Let us write a stress tensor, T, the normal of a fault plane, $\hat{n}$ and the vector $\hat{b}$ orthogonal to slip:

$$T = \begin{pmatrix} \sigma_{11} & \sigma_{12} & \sigma_{13} \\ \sigma_{12} & \sigma_{22} & \sigma_{23} \\ \sigma_{13} & \sigma_{23} & \sigma_{33} \end{pmatrix} \tag{4.115}$$

$$\hat{n} = \begin{pmatrix} n_1 \\ n_2 \\ n_3 \end{pmatrix} \tag{4.116}$$

$$\hat{b} = \begin{pmatrix} b_1 \\ b_2 \\ b_3 \end{pmatrix} \tag{4.117}$$

as usual, circumflexes indicate unit vectors and $\hat{b}$ is the vector $\hat{u}_i$ used in the developments of Section 4.2.1 (i.e. following the notations of Carey, 1976, see Fig. 4.8).

T is solution of the problem (i.e. shear stress is parallel to slip but not necessarily of same sense) if:

$$\hat{b}.(T\hat{n}) = 0 \tag{4.118}$$

which is equivalent to Eq. (4.16), the equation proposed originally by Carey and Brunier (1974).

Combining Eqs (4.115)–(4.118) we find:

$$\sum_{i=1}^{3}\sum_{j=1}^{3} b_i \sigma_{ij} n_j = 0 \tag{4.119}$$

Developing and re-arranging Eq. (4.119) leads to:

$$\begin{aligned} b_1 n_1 \sigma_{11} + b_2 n_2 \sigma_{22} + b_3 n_3 \sigma_{33} \\ + (b_1 n_2 + b_2 n_1)\sigma_{12} + (b_2 n_3 + b_3 n_2)\sigma_{23} + (b_3 n_1 + b_1 n_3)\sigma_{13} = 0 \end{aligned} \tag{4.120}$$

In order to clarify the geometrical meaning of Eq. (4.120), let us examine an equivalent equation in the 3D space:

$$ax + by + cz = 0 \tag{4.121}$$

where a, b and c are the coefficients of the equation and x, y and z its three variables or unknowns.

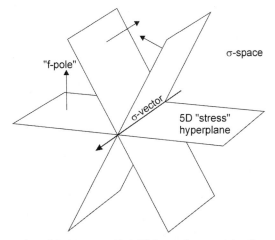

FIG. 4.18 Simplified representation of Fry's σ-space. Each 5D hyperplane contains the collection of stress tensors explaining the fault slip datum represented by its normal (i.e. 'f-pole'). As such the intersection between all the hyperplanes, i.e. σ-vector, represents the solution of the fault slip inversion problem (see text for details). Note that for the sake of clearness only three fault slip data are depicted albeit the problem requires at least four input data.

Eq. (4.121) represents a 2D plane involving the origin of the 3D space and whose normal is the vector $(a, b, c)^T$. By analogy, Eq. (4.120) represents a 5D hyperplane through the origin of the 6D σ-space, and its normal is the vector containing the coefficients of the equation, that is $(b_1 n_1, b_2 n_2, b_3 n_3, b_1 n_2 + b_2 n_1, b_2 n_3 + b_3 n_2, b_3 n_1 + b_1 n_3)^T$. Thus, the stress tensors honouring Eq. (4.118) form a 5D hyperplane in the σ-space, whose normal (i.e. the 'f-pole', Fry, 1999) represents the fault datum (Fig. 4.18). Therefore, the numerical fault slip inversion problem (see Section 4.2.2) becomes a geometrical one, which consists in finding the intersection between the 5D hyperplanes, namely the 6D σ-vector (Fig. 4.18):

$$\sigma - \text{vector} = \begin{pmatrix} \sigma_{11} \\ \sigma_{22} \\ \sigma_{33} \\ \sigma_{12} \\ \sigma_{23} \\ \sigma_{13} \end{pmatrix} \tag{4.122}$$

(see also Eq. (4.58)).

However, fault slip data can only constrain a reduced form of the stress tensor and, thus, it exists an infinity of solutions (see demonstration in Section 4.2.1.c). Rigorously speaking the intersection between the hyperplanes associated with the fault slip dataset is a plane containing an infinity of σ-vectors, each of them honouring the problem equally.

To be complete, the scheme developed above requires reduction of the stress tensor in a similar manner as in Eq. (4.18). The reduction in dimension (i.e. from 6D to 4D) may be achieved by selecting convenient values for α and β in Eq. (4.18), as discussed in Section 4.1.3, or, alternatively, by introducing a set of normalisations such as:

$$\sigma_{11} + \sigma_{22} + \sigma_{33} = 0 \tag{4.123}$$

and

$$\sigma_{11}^2 + \sigma_{22}^2 + \sigma_{33}^2 + \sigma_{12}^2 + \sigma_{23}^2 + \sigma_{13}^2 = 0 \qquad (4.124)$$

Eq. (4.123) defines the stress tensor solution of the problem as a deviator. The assumption is supported by the fact that it is always possible to find a deviator in the space of solutions (see demonstration in Section 4.2.1.c). This normalisation has been employed often in fault slip inversion methods (e.g. Carey, 1976, 1979; Etchecopar et al., 1981; Angelier et al., 1982). Its geometrical meaning is straightforward: the collection of σ-vectors solutions of the problem belong to the hyperplane of stress deviators.

The second normalisation (i.e. Eq. (4.124)), originally introduced by Fry (1999), is particularly elegant. It represents the projection of the σ-space on a unit 6D hypersphere. Thus, the intersection of this hypersphere by a hyperplane is a great hypercircle (Fig. 4.19A), analogous to great circles on 3D spheres. The pole on the unit hypersphere located at the intersection of hyperplanes, representing homogeneous fault slip data, is therefore the σ-vector solution of the fault slip inversion problem. Alternatively, the σ-vector can be determined as the pole of the girdle containing the f-poles. In brief, the fault slip inversion problem becomes a simple interpolation problem in the σ-space similar to a traditional 3D one (see Fry, 1999, 2001 for details).

Obviously, the problem becomes pleasantly linear if only the direction of the stria is considered, as it is the case in Eq. (4.118). Fry (2001) showed nevertheless that more complex nonlinear criteria are easily added as further geometrical constraints to the inversion problem (Fig. 4.19). For instance, the traditional criterion on slip sense as reformulated by Fry (2001), where $\hat{s}$ represents unit slip vector,

$$\hat{s} \cdot (T\hat{n}) > 0 \qquad (4.125)$$

corresponds geometrically to reduction of half the solution space determined from Eq. (4.118). Adding friction (see Section 4.2.4) reduces further the geometrical extend of the solutions (Fig. 4.19C).

Finally, a major outcome of Fry's analysis is the demonstration that fault slip data with undetermined senses are more valuable to constrain the reduced stress tensor than previously thought, a result later confirmed by Sato (2006). The mixing of fault slip data with variable quality is also facilitated by Fry's approach (Fig. 4.19C).

b) Modification of the concept of σ-space by Sato and Yamaji (2006b)

Sato and Yamaji (2006b) explored mathematically and simplified Fry's concept of σ-space. The modifications introduced by Sato and Yamaji rest fundamentally on the redefinition of the normalisations.

Let us consider the general expression of a symmetric second-rank tensor

$$X = \begin{pmatrix} X_{11} & X_{12} & X_{13} \\ X_{12} & X_{22} & X_{23} \\ X_{13} & X_{23} & X_{33} \end{pmatrix} \qquad (4.126)$$

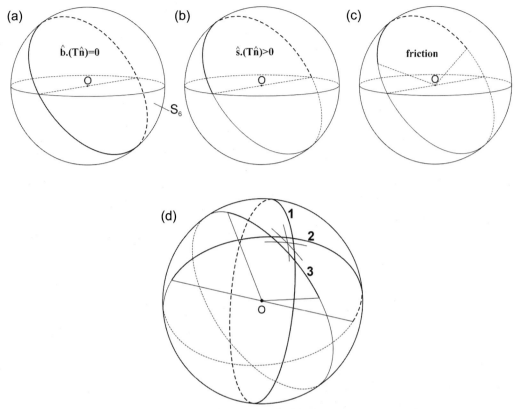

FIG. 4.19 Schematic representation of the 6D σ-space of Fry (1999, 2001). S_6 is the hypersphere corresponding to the normalisation given by Eq. (4.124). (A) The great hypercircle, intersection of S_6 by an hyperplane representing a fault slip datum, indicates the collection of reduced stress tensors honouring the datum, when shear sense is ignored (i.e. Eq. (4.118)). (B) If shear sense is taken into account (i.e. Eq. (4.125)) only half of the great hypercircle represents valid solutions. (C) If friction is introduced as additional constraint the solution domain is further restricted (see text and Fry, 2001 for details). (D) Determination of the best fit reduced stress tensor (i.e. the interpolated centre of the spherical triangle) using fault slip data with different degrees of acuracy: 1—only stria direction is known, 2—with known shear sense and 3—with known shear sense and friction.

and let us associate to it the 6D vector

$$\vec{F}_6 = \begin{pmatrix} \dfrac{X_{11}}{\sqrt{2}} \\ \dfrac{X_{22}}{\sqrt{2}} \\ \dfrac{X_{33}}{\sqrt{2}} \\ X_{12} \\ X_{23} \\ X_{13} \end{pmatrix} \tag{4.127}$$

Eq. (4.127) introduces a modification of the original definition of Fry's σ-space, the diagonal terms of the tensor are now divided by the factor $\sqrt{2}$. The use of such a weight factor is traditional in plasticity theory and its interest will appear clearly in the following.

The σ-space as redefined by Sato and Yamaji (2006b) accepts for normalisations:

$$J_I = X_{11} + X_{22} + X_{33} = 0 \tag{4.128}$$

$$J_{II} = \frac{1}{2}\left(X_{11}^2 + X_{22}^2 + X_{33}^2\right) + X_{12}^2 + X_{23}^2 + X_{13}^2 = 1 \tag{4.129}$$

where J_I and J_{II} are the first and second tensor invariants, respectively.

The first normalisation indicates that the tensor is a deviator, as already stated previously, and the second one applies only to deviators. Comparing the above equations with Eqs (4.123) and (4.124), only the second normalisation differs from the scheme of Fry (1999). Normalising invariants instead of an expression like the one of Eq. (4.124) is advantageous, because the principal values of the normalised tensor become independent of the choice of the coordinate system.

Let us re-write Eq. (4.127):

$$\vec{F}_6 = \begin{pmatrix} x_{11} \\ x_{22} \\ x_{33} \\ x_{12} \\ x_{23} \\ x_{13} \end{pmatrix} \tag{4.130}$$

Eqs (4.128) and (4.129) become, respectively:

$$x_{11} + x_{22} + x_{33} = 0 \tag{4.131}$$

and

$$x_{11}^2 + x_{22}^2 + x_{33}^2 + x_{12}^2 + x_{23}^2 + x_{13}^2 = 1 \tag{4.132}$$

The two equations take forms similar to the normalisations of Fry (1999) and bear the meaning already discussed in Section 4.2.7.a, that is, Eq. (4.131) represents the 5D hyperplane of deviators and Eq. (4.132) a 6D unit hypersphere, S_6, around the origin. The two conditions together indicate the section of the hypersphere by the hyperplane, which by definition is a 5D unit hypersphere, S_5 (note that the great hypercircle of Fry, mentioned previously, is the intersection between S_5 and S_6).

As consequence of the adopted normalisations the end point of the 6D vector $\vec{F}_6$ is on S_5, thus $\vec{F}_6$ may be conveniently replaced by its 5D counterpart, $\vec{F}_5$. The transformation consists in a simple rotation of the coordinate system, where one of the coordinate axes is set perpendicular to the hyperplane of deviators (see Appendix A in Sato and Yamaji, 2006b for details). Noteworthy, $\vec{F}_6$ and $\vec{F}_5$ represent the same geometrical object! In brief, the formalism of Sato and Yamaji (2006b) simplifies the original 6D problem of Fry (1999, 2001) into a 5D problem.

It is therefore possible to replace the 6D σ-vector of Fry (1999) by a 5D unit vector of stress, named thereafter $\vec{\sigma}^{[5]}$, without any loss of information. In addition, it is also possible to write the expressions of two vectors describing the fault slip datum in the 5D parameter space. The analytical development does not present any particular difficulty (see pp. 960–961 in Sato and Yamaji, 2006b) and leads to the formulations of $\vec{\varepsilon}^{[5]}$, a unit vector representing a strain tensor associated with fault slip, and of $\vec{\varepsilon}'^{[5]}$, a second unit vector representing a virtual strain tensor, that is, representing an imaged displacement perpendicular to fault slip (i.e. parallel to $\hat{b}$). As such, the two vectors are orthogonal in the 5D parameter space:

$$\vec{\varepsilon}^{[5]} \cdot \vec{\varepsilon}'^{[5]} = 0 \tag{4.133}$$

Sato and Yamaji (2006b) demonstrated that the work done on the rocks during faulting may be written as the dot product of the 5D stress and strain vectors:

$$W = \vec{\varepsilon}^{[5]} \cdot \vec{\sigma}^{[5]} \tag{4.134}$$

Furthermore, the main Wallace–Bott hypothesis implies that energy dissipation related to faulting is positive, hence:

$$\vec{\varepsilon}^{[5]} \cdot \vec{\sigma}^{[5]} > 0 \tag{4.135}$$

Similarly the expression of virtual work, which is nil, may be derived:

$$W' = \vec{\varepsilon}'^{[5]} \cdot \vec{\sigma}^{[5]} = 0 \tag{4.136}$$

It can be shown that Eqs (4.136) and (4.135) are strictly equivalent to Eqs (4.118) and (4.125), that is, the equations used by Fry (1999, 2001) to formulate the main Wallace–Bott hypothesis. Eqs (4.135) and (4.136) are new formulations of the main Wallace–Bott hypothesis based on energy balance considerations. In the 5D parameter space $\vec{\sigma}^{[5]}$ is therefore constrained to be perpendicular to $\vec{\varepsilon}'^{[5]}$ and to make an acute angle with $\vec{\varepsilon}^{[5]}$, thus to be located on a specific half great hypercircle of S_5 (Fig. 4.20A).

The main advantage of the geometrical formulation of the fault slip inversion problem by Sato and Yamaji (2006b) resides in the straightforward derivation of relevant quantities. The misfit angle separating measured stria and computed shear stress vector corresponds to the angular deviation of $\vec{\sigma}^{[5]}$ from the hyperplane perpendicular to $\vec{\varepsilon}'^{[5]}$ (Fig. 4.20B) and is given by:

$$\alpha = \tan^{-1}\left(\frac{\left|\vec{\varepsilon}'^{[5]} \cdot \vec{\sigma}^{[5]}\right|}{\vec{\varepsilon}^{[5]} \cdot \vec{\sigma}^{[5]}}\right) \quad \text{with } 0° \leq \alpha \leq 180° \tag{4.137}$$

Another quantity of interest easily deduced from the new formulation of the σ-space is the *stress difference*, D, originally introduced by Orife and Lisle (2003). Let us consider the difference between two reduced stress tensors: $T_D = T_A - T_B$. The magnitude of the

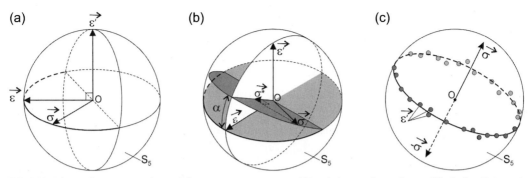

FIG. 4.20 Schematic representations of the parameter space as a 5D unit hypersphere, S_5, modified after Sato and Yamaji (2006b). (A) Geometrical relationships between $\vec{\sigma}$ (5D stress vector), $\vec{\varepsilon}$ (5D strain vector) and $\vec{\varepsilon}\,'$ (5D virtual strain vector), where $\vec{\varepsilon}$ and $\vec{\varepsilon}\,'$ represent the fault slip datum. According to Eqs (4.135) and (4.136) $\vec{\sigma}$ forms an acute angle with $\vec{\varepsilon}$ and is perpendicular to $\vec{\varepsilon}\,'$, thus the end point of $\vec{\sigma}$ is constrained to remain on the bold portion of the depicted great hypercircle. (B) Definition of the angular misfit, α, between stria and shear stress in the parameter space. The angle α is found as the deviation of the stress vector with respect to the hyperplane supporting the locus of the solutions (i.e. the great hypercircle depicted in (A)). $\vec{\sigma}\,^*$ represents the orthogonal projection of $\vec{\sigma}$ on the hyperplane ($\vec{\varepsilon}$, $\vec{\varepsilon}\,'$). (C) Determination of the 5D stress vectors honouring a collection of fault slip data, which are represented by $\vec{\varepsilon}\,'$ poles on S_5. $\vec{\sigma}$ is constrained to be perpendicular to the average hyperplane trough the $\vec{\varepsilon}\,'$ poles. The present condition corresponds to Eq. (4.136); therefore, two opposite vectors honour the problem. Additional consideration of shear senses helps to discriminate between both. See text for details. Note that the superscript [5] used in text has been omitted for the sake of clarity.

octahedral shear stress of T_D is defined as the stress difference, which varies from $D=0$ (i.e. $T_A=T_B$) to $D=2$ (i.e. $T_A=-T_B$). The latter value describes 'opposite tensors' according to the definition of Nemcok and Lisle (1995), i.e. T_A and T_B interchange the orientations of their σ_1 and σ_3 axes and their respective shape ratio values are Φ and $1-\Phi$. A simple mathematical development (see Appendix C in Sato and Yamaji, 2006b) demonstrates that:

$$D=\sqrt{J_{II}(T_D)} \qquad (4.138)$$

In the 5D parameter space, D represents an Euclidian metric measuring the distance between the end points of two $\vec{\sigma}\,^{[5]}$ vectors. If the two end points coincide, $D=0$ and the two stress vectors coincide (i.e. the two reduced stress tensors are identical). If the two stress vectors, which are unit vectors, point in opposite directions, $D=2$ and the reduced stress tensors they represent are 'opposite' to each other.

Finally, the solving of the fault slip inversion problem is divided in two parts, a linear an a nonlinear one, as originally proposed by Fry (1999). The linear part corresponds to the solving of Eq. (4.136) (i.e. Eq. (4.118)) for $N>3$ fault slip data. In turn, the solving of Eq. (4.136) is equivalent to searching for the normal of the hyperplane defined by the poles of the $\vec{\varepsilon}\,'^{[5]}$ vectors associated to the fault slip data (Fig. 4.20C). To these aims Fry (1999) and Sato and Yamaji (2006b) employed the so-called 'eigenvector problem'. The solution of the linear part of the problem furnishes two opposite stress tensors, the nonlinear part, corresponding to Eq. (4.135) (i.e. Eq. (4.125)), consists in checking shear senses to identify the stress vector (i.e. reduced stress tensor) honouring the problem.

4.3 Right dihedra method

4.3.1 Focal mechanisms of earthquakes

The right dihedra method stems from the classical determination of *earthquake focal mechanisms* in seismology (see e.g. Aki and Richards, 2009, pp. 76–82 for detailed mathematical treatment). Consider slip along a fault producing an earthquake of sufficient magnitude (Fig. 4.21A). A fault is finite in length and displacement of the fault blocks is therefore arrested at its tips, resulting at the instant of fault rupture in zones under compression or extension, themselves distributed around the fault in a symmetrical pattern.

The latter phenomenon is evidenced by the motion of the ground consecutive to the first *P*-wave arrivals, which is directed upwards for the two compressional quadrants and downwards for the tensional ones. When the number and the spread of seismic stations is large enough and if most of the recorded first P-wave motions can be easily read, it is possible to identify accurately these quadrants and the two planes of symmetry separating them (i.e. the *nodal planes*) by plotting on a net the relative locations of the seismic stations and the corresponding polarities of the first P-wave motions (Fig. 4.21A). One of the nodal planes corresponds to the actual fault, the second one (i.e. the *auxiliary plane*) is a virtual plane of symmetry, orthogonal to the slip vector. Therefore, fault slip is fully determined by the construction of the focal mechanism of the earthquake. To note, if the fault does not produce rupture of the ground surface, it is in general difficult to find which one of the two nodal planes is the true fault.

Traditionally, tensional quadrants are left blank and compressional ones are indicated in dark grey or red. Such representations of earthquake focal mechanisms are often called *beachballs* for obvious reasons. Three symmetry axes are also customarily defined: the *P* and *T* axes for the tensional and compressional quadrants, respectively, and the *B* axis parallel to the intersection of the two nodal planes. The letters '*P*' (i.e. pressure or compression) and '*T*' (i.e. tension) originally selected to define the symmetry axes of the tensional and compressional quadrants, respectively, are undoubtedly confusing and a common source of mistakes in the literature. Rigorously speaking *P*, *B* and *T* axes are symmetry axes of focal mechanisms and nothing else.

4.3.2 Principle of the determination of the stress axes

In contrast to the stress inversion methods previously described in Section 4.2, the *right dihedra method* (Angelier and Mechler, 1977) is based on geometrical constructions, and these involve various focal mechanisms. A right dihedron is formed by two orthogonal planes, such as the two nodal planes of a focal mechanism, and divides the space in two pairs of quadrants. To understand the principles of the method, let us examine the kinematics of the strike-slip fault depicted in Fig. 4.9A. Dextral strike-slip along the fault automatically implies σ_1 to be located in the tensional quadrants and σ_3 in the compressional ones, as demonstrated mathematically by McKenzie (1969). Furthermore, McKenzie showed in his analytical treatment that the principal axes of stress do not coincide, in

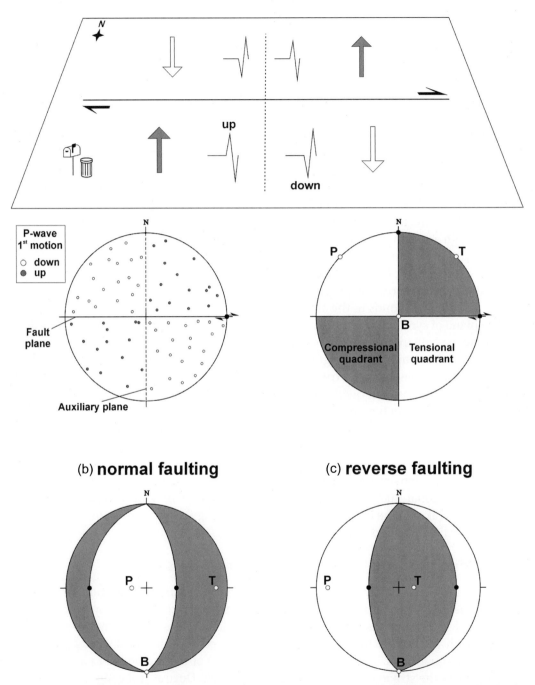

FIG. 4.21 (A) construction of the focal mechanism of an earthquake consecutive to strike-slip faulting. Compressional and tensional quadrants are separated by the two nodal planes, i.e. the fault and the auxiliary plane, and identified by the polarity, respectively, up and down, of the first P-wave motions. The auxiliary plane is the plane perpendicular to the fault slip vector. Projection of the seismic stations together with the recorded polarities of the first P-wave arrivals on a net allows for constructing the focal mechanism of the earthquake, or beachball, and for defining the orientation of the symmetry axes, P, B and T. (B) and (C) depict earthquake focal mechanisms for normal and reverse faulting cases, respectively. Note that in general it is not possible to decide which one of the two nodal planes is the actual fault plane; therefore, black dots represent for each focal mechanism the two possible fault slip vectors.

general, with the symmetry axes of the beachball (e.g. σ_1 does not coincide with P). Coincidence between principal stress and symmetry axes can only occur if the fault plane is at 45° from both σ_1 and σ_3. Such a case would be physically impossible if the fault were created during the faulting event, because it would require an internal friction coefficient equal to zero for the rocks involved in the rupture process (see Section 3.2.3 and Eq. (3.118)). The only case the two sets of axes can coincide is the fortuitous one where a discontinuity at 45° of σ_1 is reactivated in shear, and the latter occurrence is only possible when the Mohr circle crosses the friction line (see Section 3.2.2 and Fig. 3.14). In brief P, B and T axes are, a priori, rather crude proxies for the orientations of the principal axes of stress. Célérier (1988) demonstrated, however, that the maximum mismatch between symmetry and stress axes is much less dramatic than predicted by McKenzie (1969), if fault friction is taken into account (see Section 4.2.4). Nonetheless, in the absence of constrains on fault friction, the only information that can be gathered from one single beachball is that σ_1 and σ_3 are located in the tensional and compressional quadrants, respectively.

However, if fault slip happens on faults with different attitudes, their respective beachballs exhibit contrasting patterns, that is the locations and shapes of the compressional and tensional quadrants, as imaged on a net, differ from one beachball to the other (Fig. 4.22). Assuming a constant stress state in space and time, σ_1 belongs automatically

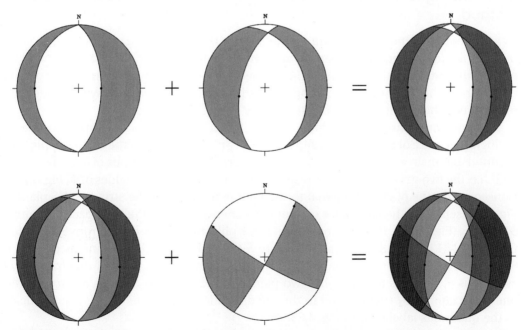

FIG. 4.22 Synthetic example of application of the right dihedra method to determine the locations of the principal axes of stress. For each individual beachball, σ_1 and σ_3 are within the tensional (*white*) and the compressional (*grey*) quadrants, respectively. The superposition of various beachballs helps to refine stress axes locations, σ_1 being contained in the remaining white areas and σ_3 being located in the darkest areas. Let us remark that the solution is particularly improved when contrasting fault orientations and types are involved in the analysis. Black dots indicate potential fault slip vectors.

to all the tensional quadrants and σ_3 is contained in all the compressional quadrants. Therefore, the areas enclosing, respectively, the two axes of principal stress can be substantially reduced in superposing all the beachballs. The procedure is equivalent to eliminating overlaps between tensional and compressional zones, following the reasoning that σ_1 or, conversely, σ_3 cannot be at the same time in the compressional quadrant of a given beachball and the tensional quadrant of the following one. In the example given in Fig. 4.22, these areas excluded from the solution appear in light to medium grey, the white and darkest grey areas being the loci of σ_1 and σ_3, respectively. Noteworthy, the right dihedra method is not restricted to the analysis of focal mechanisms of earthquakes but can also be applied to fault slips measured in the field, the construction of the missing auxiliary planes (i.e. the planes perpendicular to the faults and their respective striae) being a trivial task. The only difference is that in the latter case the actual fault planes are known. Compared to mathematical inversion methods, the right dihedra method cannot resolve the shape ratio, Φ, in most cases (however, Delvaux and Sperner, 2003 propose a statistical method to obtain reasonable estimates of Φ). Estimating the Φ value is nevertheless possible for collections of fault slips reflecting unambiguously extreme values of the shape ratio, i.e. cases of uniaxial compression or extension.

4.3.3 The right trihedra method: An extension of the right dihedra method

As a refinement of the right dihedra method, Lisle (1987, 1988, 1992) introduced the *right trihedra method* to reduce further the areas hosting, respectively, the maximum and the minimum principal axis of stress. A right trihedron is defined by three mutually orthogonal planes. The right trihedra method takes advantage of the inherent symmetries existing between the orientations of the stress axes and the slip vector.

As shown previously (see Section 4.1.2 Fig. 4.6), it is always possible to determine the cyclographic trace of possible fault slips, and every pole (i.e. stria) of the cyclographic trace is bound to remain within the arc defined by the orthogonal projections of σ_1 and σ_3 on the fault (i.e. the two striae S' and S''' corresponding to the two extreme values of Φ, Fig. 4.23A). Consequently, the plane orthogonal to the fault and including the stria (i.e. NS plane in Fig. 4.23) separates the half-space containing S', hence σ_1, from the half space containing S'', thus σ_3 (Fig. 4.23B). Furthermore S' and S'' are obligatory at 90° or less from S. Therefore the plane NU (Fig. 4.23C), perpendicular to S (indeed the auxiliary plane as defined in Section 4.3.1), together with plane NS divide the space in two pairs of quadrants such as S', thus σ_1, belongs to one of the two pairs of quadrants and S'', thus σ_3, to the other one. For example, σ_1 is contained in the quadrants labelled A and σ_3 falls automatically in the quadrants labelled B in Fig. 4.23.

A synthetic example of application of the right trihedra method is given in Fig. 4.24. Let us consider a collection of focal mechanisms of earthquakes involving one nodal plane identified as a true fault (e.g. one fault producing rupture of the Earth's surface).

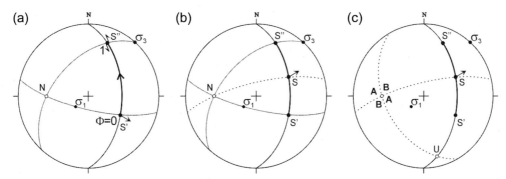

FIG. 4.23 Principles of the right trihedra method. (A) graphical construction of possible fault slips (cyclographic trace in bold) in function of Φ for a given fault and fixed orientations for the principal axes of stress (see also Section 4.1.2 and Fig. 4.6). (B) The NS plane perpendicular to the fault and parallel to the stria, S, separates two half-spaces containing, respectively, S', thus σ_1, and S'', thus σ_3. (C) NS and NU (i.e. the auxiliary plane) divide the space in two pairs of quadrants, arbitrarily labelled A and B. Each pair of quadrants contains either σ_1 or σ_3 exclusively (in the present example, σ_1 is in A and σ_3 in B, see text for details).

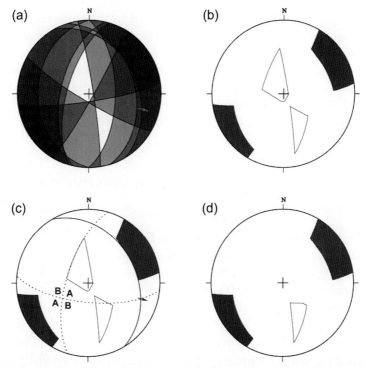

FIG. 4.24 Application of the right trihedra method to a synthetic case. (A) collection of beachballs, one fault plane being unambiguously identified, and (B) solution derived using the right dihedra method (white and dark areas contain σ_1 and σ_3, respectively). (C) refinement of the solution given in (B) using the right trihedra method after the knowledge of one fault plane. The dark areas containing σ_3 are located exclusively in quadrants A, therefore σ_1 has to fall in quadrants B and (D) the white areas within quadrants A are automatically excluded from the solution, leading to significant improvement in the determination of σ_1.

The method, evidently, can be applied to fault slip data measured in the field as well. Preliminary analysis using the right dihedra method leads to the determination of the areas enclosing the respective locations of σ_1 and σ_3 (Fig. 4.24A and B). After construction of the right trihedron based on the identified fault plane, its auxiliary plane and the plane perpendicular to both (thus containing the stria), the space is divided in two pairs of quadrants, each pair containing exclusively either σ_1 or σ_3. The two quadrants labelled A in Fig. 4.24C involve partly the white areas (i.e. potential locations for σ_1) and completely the two dark areas (i.e. potential locations for σ_3). In turn, the quadrants labelled B contain only a portion of one of the two white areas and no dark areas. In conclusion, σ_3 cannot be located in B but in A and, therefore, σ_1 has to plot in B. Thus, the white areas contained in A are automatically excluded from the range of solutions, leading to substantial refinement of the location of σ_1 in the present case (Fig. 4.24D).

The possibility of testing which one of the two nodal planes of a beachball is the actual fault represents an additional advantage of the right trihedra method. Once the loci of the axes of principal stress are defined, each nodal plane can be tested by means of constructing its corresponding right trihedron. The nodal plane is potentially the true fault if the pairs of quadrants are consistent with the locations of the areas containing the stress axes.

The two nodal planes of one of the beachballs used for the stress determination devised in Fig. 4.24 are tested in the example depicted in Fig. 4.25. Assuming that the west-dipping nodal plane is the true fault results in a geometrical configuration where both white and dark areas are located in A (Fig. 4.25B). In other words the result is inconsistent. In turn, assuming that the east-dipping nodal plane corresponds to the real fault, leads to consistent distribution of the areas containing the stress axes, the white area being located in B and the two dark areas in A (Fig. 4.25C). Therefore, the analysis suggests the east-dipping nodal plane as being the true fault.

4.4 Validity and limits of fault slip inversion methods

4.4.1 To what extent is the main Wallace–Bott hypothesis valid?

The whole edifice of fault slip inversion methods holds on the main Wallace–Bott hypothesis (i.e. fault slip occurs parallel to the maximum shear stress, see Section 4.1.2), which is the only one making a firm link between field observations (i.e. slickenlines) and mathematical objects (i.e. shear stresses). It is thus scientifically healthy to question the extent to which this hypothesis is valid and to try to challenge it merciless.

First, we follow Lisle (2013) and examine the variation of shear stress magnitudes around the maximum shear stress vector resolved on the plane, that is, around the orthogonal projection of the stress vector, $\vec{\sigma}$ (Fig. 4.8). Let us consider the shear stress generated by oblique projection of the stress vector. Shear stress magnitudes decrease from a maximum value, τ_{max}, obtained for the orthogonal projection of $\vec{\sigma}$ to zero, when it is projected at 90° of the direction of the maximum shear stress. It is interesting to note that shear stress magnitudes decrease slowly away from the direction of the maximum shear stress.

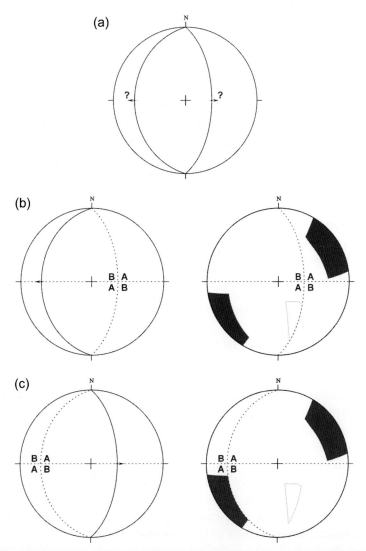

FIG. 4.25 Use of the right trihedra method to determine (A) which one of the two nodal planes is the actual fault. (B) Assuming the west-dipping nodal plane to be the true fault leads to an inconsistent result, for which all dark and white areas are contained in the same quadrants (A in the present case). (C) Assuming the east-dipping nodal plane to be the true fault results in a consistent distribution of the areas, the white area belonging to B and the dark ones to A. Thus, the latter nodal plane is a priori the actual fault. The nodal planes correspond to one of the beachballs depicted in Fig. 4.24A, the solution for the orientations of the stress axes is the one indicated in Fig. 4.24D.

It is straightforward to show that shear stress magnitudes remain greater than 0.9 τ_{max}, if the shear stress vector deviates less than ±26° from the direction of the orthogonal projection of $\vec{\sigma}$. Summarising, this simple reasoning suggests that even modest perturbations to the background stress-slip model may produce substantial deviation of fault slip from the ideal orientation of the maximum shear stress vector. The fact is acknowledged by most fault slip inversion methods, which tolerate relatively large angular deviations between measured slip and computed shear stress (i.e. close to $\pm20°$ typically).

One of the most critical sources of perturbations might be interaction between faults, a phenomenon that, erroneously, Arthaud (1969) assumed to be strong enough to hamper any paleostress reconstruction. Fault slip alters locally the stress field (see Fig. 7.10) and may consequently disturb slip on neighbouring faults. The extent of the perturbations were difficult to access before the advent of innovative numerical methods, capable to compute 3D mechanical models of discontinuous media involving fault block deformation, and before enhanced computer performance by the end of the 20th century.

The pioneering numerical studies of Dupin et al. (1993) and Pollard et al. (1993) showed that slip perturbations were spatially restricted to fault tips and, in case of intersecting faults, to the vicinity of their intersection line. The remarkable outcome of these modelling studies was the demonstration that the vast majority of the slips computed on the fault surfaces were in good agreement with Wallace–Bott predictions, and deviations between modelled and Wallace–Bott slips much less than $\pm20°$, in general. Maerten (2000) and Pascal et al. (2002) extended these early studies in modelling more complex cases of intersecting faults. Their works confirmed the previous findings, albeit the uniaxial stress state adopted by Maerten (2000) had the tendency to enhance slip perturbations (see discussion in Pascal, 2002).

The number of cases modelled by these authors was, however, limited by the technology at their time. Systematic numerical exploration of the problem became possible only recently with the development of an optimised Boundary Element Method (Maerten, 2010 and Section 7.3 for an overview of the method) and new generations of personal computers. Lejri (2015) and Lejri et al. (2015, 2017) modelled numerous configurations of intersecting faults while exploring the whole range of Andersonian stress states and the influence of mechanical parameters and, in particular, of friction. This group of authors concluded that in most cases the main Wallace–Bott hypothesis was validated. They also showed that standard to relatively high friction coefficients reduced deviation between modelled and Wallace–Bott slips. The main lesson learned from all the previously cited modelling studies is that fault slips consistent with the main Wallace–Bott hypothesis dominate statistically, even in cases of intersecting faults.

However, all these previous numerical models assumed isotropic frictional properties along the fault planes. For example, preexisting corrugations (i.e. grooves) with modest wavelength but significant amplitude may promote anisotropic friction, friction being minimum parallel to the corrugations but maximum perpendicular to them (Lisle, 2013). For such particular fault plane geometry, slip is expected to be deviated towards the direction of the corrugations. Lisle (2013) suggested that slip deviation could reach up to 30° but anticipated lower deviations at depth, because of the mechanical smoothing of re-activated fault planes with increasing stress.

The assumption of planar faults is obviously an important issue, as departures from planar geometries may have other consequences than friction anisotropy. Pronounced variations in fault strike and/or dip have often been invoked as serious challenges for the

Wallace–Bott hypothesis (e.g. Lisle, 2013). However, most of the criticisms emerged from simplified kinematic analyses, which assumed rigid fault blocks (e.g. Nieto-Samaniego and Alaniz-Alvarez, 1997). Lejri et al. (2015) modelled slip on faults with long wavelengths corrugations of varying amplitudes. Maximum misfits of up to 35° were found for few particular cases and zero friction, that is, for end-member cases. In general, deviations between modelled and Wallace–Bott slips were between 0° and 15°–20°. Deformation of fault blocks in geomechanical models accommodates part of the kinematic incompatibilities, which otherwise are solved by uniform block motions in rigid block models, and thus modelled slips remain closer to those predicted by Wallace–Bott hypothesis. In conclusion, it seems that departure of faults from planar geometries does not affect the validity of the main Wallace–Bott hypothesis, at least in the cases modelled by Lejri et al. (2015).

Similarly, kinematic analyses assuming rigid blocks have predicted strong deviations from the predictions of the Wallace–Bott model in case of intersecting faults (Nieto-Samaniego and Alaniz-Alvarez, 1997). The 'wedge configuration', where two faults with mutually orthogonal strikes intersect, is a typically studied case. Kinematic models predict that slip on both faults parallels automatically the intersection line. However, the numerical results of Pascal et al. (2002) suggest that fault slips parallel the intersection line only in its close vicinity. Fault slips rotate gradually away from the intersection line to adopt directions in good agreement with Wallace–Bott hypothesis. Once again, kinematic analyses appear to deliver rather poor predictions of fault slips in assuming rigid blocks.

In conclusion, the main Wallace–Bott hypothesis is definitively robust. To date, all the published geomechanical numerical modelling tests suggest that fault slips are coherent with the hypothesis in most cases. This latter outcome implies that fault slips measured in the field should statistically comply with the main Wallace–Bott hypothesis. The modelling results point however to a minority of cases where the hypothesis breaks down (see in particular the results of Lejri et al., 2017). Nevertheless, this minority of cases may be identified and avoided in the field or, if not, it is anticipated to generate outliers that are easily removed from the database.

4.4.2 Stress, strain, strain rate or nothing else than displacement data?

The actual meaning of the mathematical object reconstructed by fault slip inversion methods is certainly the most debated issue and prompts periodic and sometimes animated discussions in scientific meetings and journals (e.g. Marrett and Allmendinger, 1990; Twiss and Unruh, 1998; Gapais et al., 2000; Lacombe, 2012; Simón, 2019).

As early as the prime infancy of fault slip inversion methods, Arthaud (1969) discussed in depth the so-called 'kinematic' and 'dynamic' paradigms (Simón, 2019) to conclude that the results of his graphical method were more safely interpreted in terms of principal axes of finite strain (i.e. kinematic paradigm) than in terms of principal axes of stress (i.e. dynamic paradigm). Arthaud (1969) based his opinion on the assumption that fault interactions could invalidate the dynamic interpretation of the results. However, the latter assumption has been seriously challenged by numerical modelling studies conducted decades after Arthaud's publication (see discussion in Section 4.4.1).

Marrett and Allmendinger (1990) favoured as well kinematic interpretations of fault slip inversion. In particular, they proposed to reconstruct finite strain from moment tensor summation. Their approach required, however, the weighting of the fault slip data, which in turn required knowledge of net displacement on every fault. A major criticism of the method is that such knowledge is difficult and most often impossible to acquire in standard field conditions (e.g. Simón, 2019).

Gapais et al. (2000) advanced another view on the problem. They compared orientations of principal axes computed by inversion of fault slips, at different sites, with displacement vectors obtained from regional-scale numerical restorations in map view. The good agreement between the two sets of results let Gapais et al. (2000) conclude that fault slip inversion informed mainly on the orientation of local finite displacement (though they did not rule out completely that it could inform on local stress states as well). Such an opinion may be justified in the context of complex deformation, involving e.g. block rotations or successive fault generations, where the conditions fixed by Wallace–Bott hypotheses do not hold. It does not invalidate the dynamic paradigm in general. Additionally, one may wonder whether local displacement/deformation may be recovered at all if the deformation is too much complex.

Building on the findings of Twiss et al. (1991, 1993), Twiss and Unruh (1998) developed a thorough discussion on the kinematic vs. dynamic question. They advanced as foundations of their reasoning the following arguments: (1) fault slips represent displacements and (2) all fault slips within a given volume add to create deformation of the volume (as already pointed out by Arthaud, 1969 and Carey, 1976). Hence, the kinematic interpretation was expected to be more natural than the dynamic one. In addition, they noted that deformation through slip along numerous fault planes within a quasi-continuous volume was equivalent to cataclastic flow, a nonlinear deformation process.

Assuming fault slip parallel to maximum rate of shear strain on the fault plane (Marrett and Allmendinger, 1990), Twiss and Unruh (1998) proposed that fault slip inversion resulted in the determination of the orientations of the principal axes of strain rate, the shape of the strain rate ellipsoid and a fifth parameter, namely, the relative vorticity (i.e. an independent parameter describing fault block rotation, Twiss et al., 1991, 1993). The dynamic interpretation could remain valid in case of parallelism between principal axes of strain rate and principal axes of stress and shape conservation between the two corresponding ellipsoids, as it was assumed by early workers (e.g. Carey, 1976). Neglecting vorticity, such a particular situation was judged theoretically possible if (1) the material were isotropic (i.e. parallelism condition) and (2) the relationship between stress and strain rate were linear (i.e. conservation of ellipsoid shape). However, Twiss and Unruh (1998) pointed out that the constitutive equation for cataclastic flow, that is the relationship between stress and strain rate in this deformation mode, was expectedly complex enough to relegate the dynamic interpretation to particular cases.

The argumentation of Twiss and Unruh (1998) is definitively sound, rigorous and appears to cast serious doubts on the interpretation of fault slip inversion results as reduced stress tensors, in most cases. However, empirical data call in favour of the dynamic interpretation. For example, the results for the Late Miocene paleostress field in western Europe by Bergerat (1987), depicted in Fig. 9.1B (see Section 9.1), resemble strikingly the present-day stress field derived from in-situ stress measurements (e.g. Heidbach et al., 2018). Furthermore, Lacombe (2012) shows data from Taiwan, where Plio-Quaternary paleostresses directions, as reconstructed from inversion of fault slip data, are in excellent agreement with present-day determinations based on borehole breakout data (and with stress orientations derived from inversion of focal mechanisms of earthquakes). Lacombe (2012) states also: "Among the several thousands of paleostress reconstructions carried out worldwide using fault slip data in various tectonic settings over the last 35 years, most have proved consistent with the orientations of major structures, with the statistics of microtectonic data such as stylolithes and joints (e.g., Mattauer and Mercier, 1980; Pollard and Aydin, 1988), the significance of which as reliable stress indicators being beyond doubt, and even with paleostresses inferred from calcite twins." An opinion that appears to be shared by numerous authors, including Simón (2019). Finally, it is interesting to note that Lisle and Srivastava (2004) analysed a large number of fault slip data, previously used in paleostress reconstructions, and showed that the data were consistent with fault reactivation submitted to friction laws. In other words, their study seems to confirm the validity of the stress interpretation of fault slip inversion results.

In brief, fault slip inversion should not constrain stresses, theoretically, but empirical evidence apparently shows otherwise. How this paradox can be solved? Let us examine the background of the theory of Twiss and Unruh (1998). First, vorticity may be neglected in most cases, either because fault block rotations are merely rare (in weakly deformed terrains and for faults with minor offsets at least) or because the field geologist is capable to detect and avoid them. Until now, we only follow the recommendations of the pioneers of fault slip inversion (e.g. Carey, 1976, 1979; Angelier, 1975). Admittedly, rocks are anisotropic in most cases and, a priori, the relationship between stress and strain is not linear. The degree the characteristics of the reconstructed strain rate tensor are going to depart from those of the stress tensor is strongly dependent on the nonlinearity of the constitutive law and its sensitivity to anisotropy. Unfortunately, the constitutive law of cataclastic flow produced by multiple fault slips is unknown to date. One way to reconcile theory and empirical observations is to suppose that this law may exhibit nearly linear behaviour for most natural cases and is weakly sensitive to rock anisotropy.

Admitting that fault slip inversion returns reduced paleostress tensors, another relevant question might be: what kind of stress tensor is reconstructed? The reticence of conceiving a constant stress state during one tectonic phase, which might have lasted hundreds of thousands or even millions of years, constitutes one of the main reasons the methods have been received with much scepticism. Most authors have proposed that paleostress tensors represent 'bulk' or 'average' tensors (e.g. Angelier 1994, Lacombe, 2012). Based on examples given above (e.g. Bergerat, 1987), long-term (relative) stability in stress axes orientations appears to be a reasonable assumption. However, it seems difficult that shape ratios remain constant, at least because these are sensitive to burial depth (Rispoli and Vasseur, 1983), which is turn might vary significantly during the time span of a single tectonic phase.

Lisle et al. (2006) analysed the results of ~2000 paleostress reconstructions. They found that about 2/3 of the reconstructed reduced stress tensors were in agreement with Andersonian stress states. Thus, according to their analysis, methods assuming Andersonian conditions might be faulty in ~33% of cases. The second interesting outcome was that the average shape ratios indicated triaxial stress configurations with $\Phi \sim 0.4$. Sato and Yamaji (2006a) demonstrated that $\Phi = 0.5$ is statistically the most expected value. Lisle et al. (2006) explained their slightly lower Φ value by the effects of dominant plane strain conditions in nature and/or those of overburden pressure.

To conclude this chapter we may speculate further on the significance of paleostress tensors. Crustal stress magnitudes are not constant but fluctuate in relatively short times corresponding to seismic cycles. In-situ stress measurements deliver stress magnitudes and shape ratios, which represent snapshots during the cycle. However, because paleostresses are deduced from fault slips, they relate specifically to brittle failure conditions and not to the preceding elastic loading of the system.

Further reading

Anderson, E.M., 1951. The dynamics of faulting and dyke formation with applications to Britain. Oliver and Boyd.

Angelier, J., 1994. Fault slip analysis and paleostress reconstruction. In: Hancock, P.L. (Ed.), Continental Deformation. Pergamon, Oxford, pp. 101–120.

Bott, M.H.P., 1959. The mechanics of oblique slip faulting. Geol. Mag. 96, 109–117.

Célérier, B., Etchecopar, A., Bergerat, F., Vergely, P., Arthaud, F., Laurent, P., 2012. Inferring stress from faulting: from early concepts to inverse methods. Tectonophysics, Crustal Stresses, Fractures, and Fault Zones: The Legacy of Jacques Angelier 581, 206–219.

Ramsay, J.G., Lisle, R.J., 2000. The Techniques of Modern Structural Geology. In: Volume 3: Applications of continuum mechanics in structural geology (Session 32: Fault Slip Analysis and Stress Tensor Calculations). Academic Press, London.

Wallace, R.E., 1951. Geometry of shearing stress and relation to faulting. J. Geol. 59, 118–130.

Yamaji, A., 2007. An Introduction to Tectonophysics: Theoretical Aspects of Structural Geology (Chapter 11: Determination of Stress from Faults). Terrapub, Tokyo.

Inversion of tensile fractures

5.1 Pore fluid pressure and opening of pre-existing discontinuities

5.1.1 2D analysis of the problem

Anderson (1938, 1951) advanced the very first theory bridging orientations of tensile fractures, tectonic stress and pore pressure. He postulated that tensile fractures, particularly magmatic dykes and sills, open and potentially propagate when the pressure of the fluids they host exceeds normal stresses acting on their walls (Fig. 5.1). In its simplest form, the latter statement can be written as follows:

$$P_f \geq \sigma_n \tag{5.1}$$

where P_f is pore fluid pressure and σ_n normal stress.

Consequently, the minimum pore pressure required to open a fracture corresponds to the minimum principal stress; therefore, most tensile fractures are expected to form perpendicular to σ_3. In addition, he used the analytical solutions of Inglis (1913) to suggest that the concentration of tensile stresses ahead of the tips of a fracture assists its propagation parallel to the mean fracture plane (i.e. 'wedging'; Fig. 5.1). Based on his field observations on the different dyke systems of Scotland, he distinguished two cases: (1) tensile fractures created during intrusion, the most common case according to Anderson; and (2) pre-existing discontinuities reactivated in tension. The mathematical approach adopted by Anderson, where tensile strength is neglected, appears to be better suited for the second case.

Although Anderson did not rule out that pore pressure can also compete against the intermediate and maximum principal stress and, thus, lead to fracture opening oblique to σ_3, he did not investigate in detail these potential occurrences. Jolly and Sanderson (1997) extended the 2D analysis by Delaney et al. (1986) to determine the orientations of pre-existing discontinuities reactivated in tension as a function of pore fluid pressure and submitted to a given stress state. In the case of reactivation, negligible tensile strength is expected; therefore, Eq. (5.1) represents a reasonable first-order approximation. Uniform distribution of the orientations of the pre-existing planar discontinuities is the fundamental assumption of Jolly and Sanderson's formalism and, consequently, of tensile fracture inversion methods in general.

Let us consider the expression of the normal stress applied on a planar discontinuity given by Eq. (3.78) in Chapter 3. Combining (3.78) and (5.1), we get

$$P_f - \frac{\sigma_1 + \sigma_3}{2} - \frac{\sigma_1 - \sigma_3}{2} \cos 2\theta \geq 0 \tag{5.2}$$

Paleostress Inversion Techniques. https://doi.org/10.1016/B978-0-12-811910-5.00009-9

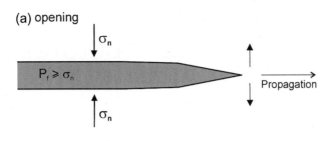

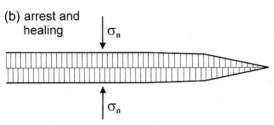

FIG. 5.1 (A) Opening of a tensile fracture when pore fluid pressure overcomes normal stress on fracture walls; propagation parallel to the fracture plane is driven by the concentration of tensile stresses beyond the tip of the fracture (Anderson, 1938, 1951). (B) The fracture opens and propagates, i.e. fracture volume increases and P_f decreases, until a new equilibrium is reached (Secor, 1965) and the fracture is arrested and potentially heals (e.g. through vein mineralisation or cooling of injected magma).

$$\frac{(P_f - \sigma_1) + (P_f - \sigma_3)}{(\sigma_1 - \sigma_3)} \geq \cos 2\theta \qquad (5.3)$$

Following Delaney et al. (1986), we introduce the *driving stress ratio* or *driving pressure ratio*

$$R = \frac{(P_f - \sigma_1) + (P_f - \sigma_3)}{(\sigma_1 - \sigma_3)} \qquad (5.4)$$

and rewrite (5.3) as

$$R \geq \cos 2\theta \qquad (5.5)$$

The driving stress ratio is a measure of the relative value of pore fluid pressure with respect to the minimum and the maximum principal stress. The orientations of the planar discontinuities reactivated in tension depend on this value. Clearly, Eq. (5.5) has no solution if $R < -1$ (i.e. $P_f < \sigma_3$), i.e. no discontinuity opens. For $R = -1$ (i.e. $P_f = \sigma_3$), only discontinuities orthogonal to σ_3 open, which is the typical case devised by Anderson and commonly described in classical textbooks. When $-1 < R < 1$, differently orientated planes open, the range of possible orientations depending on the value taken by the driving stress ratio, and for $R \geq 1$ (i.e. $P_f \geq \sigma_1$), all planes are reactivated independently of their respective orientations.

It is particularly convenient to use a Mohr diagram to visualise the role of R in the reactivation of planar discontinuities (Jolly and Sanderson, 1997 and Fig. 5.2). By definition,

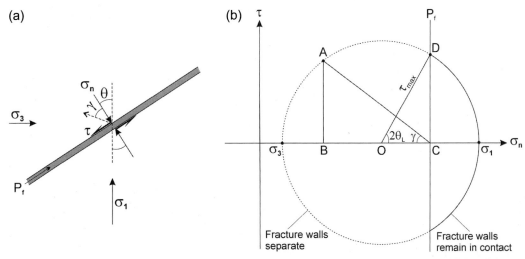

FIG. 5.2 (A) Schematic 2D representation of a planar discontinuity opening under a given stress and pore fluid pressure (P_f) exceeding normal stress on its walls. The angle γ indicates the direction of opening; other symbols follow the conventions of this book. (B) Visualisation of the orientations of the opened fractures in the 2D Mohr space; all fractures meeting the condition $P_f \geq \sigma_n$ are susceptible to open (see text for details).

the discontinuities reactivated in tension are the ones for which pore pressure exceeds normal stress, thus the ones whose poles on the Mohr circle plot left of the pore fluid pressure line (i.e. dashed portion of the Mohr circle in Fig. 5.2B). Consider the orientation of the planes for which pore pressure and normal stress are equal (i.e. the angle θ_L), the orientation of a reactivated discontinuity obeys:

$$\theta_L \leq \theta \tag{5.6}$$

$$\text{or} \quad \cos 2\theta_L \geq \cos 2\theta \tag{5.7}$$

as a quick glance at Fig. 5.2B confirms.

Let us rewrite the driving stress ratio:

$$R = \frac{P_f - \sigma_m}{\tau_{max}} \tag{5.8}$$

where σ_m is the mean stress and τ_{max} the maximum shear stress (see also Section 3.1.5):

$$\sigma_m = \frac{\sigma_1 + \sigma_3}{2} \tag{5.9}$$

$$\tau_{max} = \frac{\sigma_1 - \sigma_3}{2} \tag{5.10}$$

According to the geometrical relationships depicted in Fig. 5.2B:

$$OC = P_f - \sigma_m \tag{5.11}$$

$$OC = \tau_{max} \cos 2\theta_L \tag{5.12}$$

Thus, combining Eqs (5.8), (5.11) and (5.12):

$$R_L = \cos 2\theta_L \qquad (5.13)$$

Finally, combining Eqs (5.7) and (5.13) we find again the condition for reactivation given by Eq. (5.5).

The 2D Mohr circle approach presents the additional advantage of delivering directly opening directions. Fig. 5.2 shows that pre-existing discontinuities are expected to open as hybrid fractures in general. The direction of opening is parallel to the orientation of the effective stress vector σ' (see Section 3.2.4), conveniently defined by γ, the angle between σ' and the normal of the plane (Fig. 5.2A).

By definition (e.g. Fig. 3.4):

$$\tau = \sigma' \sin\gamma \qquad (5.14)$$

$$\sigma'_n = \sigma' \cos\gamma \qquad (5.15)$$

where

$$\sigma'_n = \sigma_n - P_f \qquad (5.16)$$

Combining Eqs (5.14)–(5.16) and multiplying the result by −1 in order to express γ as a positive value, we find:

$$\tan\gamma = \frac{\tau}{P_f - \sigma_n} \qquad (5.17)$$

Examining Fig. 5.2B, we find that the ratio on the right side of Eq. (5.17) corresponds to the ratio of AB on BC, that is:

$$\tan\gamma = \frac{AB}{BC} \qquad (5.18)$$

Thus, the opening direction of the discontinuity represented by pole A is directly given by the acute angle separating AC from the abscissa. We remark that the limiting case $\theta = \theta_L$ implies $\gamma = 90°$ or, in other words, 'opening' parallel to the fracture plane.

5.1.2 3D analysis of the problem

The extension of fracture opening in 3D was originally formalised by Baer et al. (1994) and later completed by Jolly and Sanderson (1997).

Let us consider Eq. (3.55) and combine it with (5.1):

$$P_f - l^2\sigma_1 - m^2\sigma_2 - n^2\sigma_3 \geq 0 \qquad (5.19)$$

We recall that l, m and n are the direction cosines of the unit vector normal to the planar discontinuity; as such, $l^2 + n^2 + m^2 = 1$ (see Eq. (3.43)).

Expressing n^2 in the function of l^2 and m^2 and replacing it in Eq. (5.19) gives

$$P_f - \sigma_3 - l^2(\sigma_1 - \sigma_3) - m^2(\sigma_2 - \sigma_3) \geq 0 \qquad (5.20)$$

We divide Eq. (5.20) by the differential stress, which is a positive quantity:

$$\frac{P_f - \sigma_3}{\sigma_1 - \sigma_3} - l^2 - \frac{\sigma_2 - \sigma_3}{\sigma_1 - \sigma_3} m^2 \geq 0 \qquad (5.21)$$

The driving stress ratio in 3D, R'' can be expressed as follows (Baer et al., 1994):

$$R' = \frac{P_f - \sigma_3}{\sigma_1 - \sigma_3} \qquad (5.22)$$

The term next to m in Eq. (5.21) is the shape ratio, Φ (see Eq. (3.38)); hence, Eq. (5.21) can be rewritten as

$$R' - l^2 - \Phi m^2 \geq 0 \qquad (5.23)$$

Eq. (5.23) represents the condition for tensile reactivation of pre-existing planar discontinuities in as a function of R' and Φ, i.e. as a function of relative pore pressure and stress anisotropy, for the given orientations of the principal axes of stress. The condition does not depend on actual stress magnitudes.

Let us further examine this condition. If $R' < 0$ (i.e. $P_f < \sigma_3$), Eq. (5.23) has no solution or, in other words and expectedly, no discontinuity opens. If $R' = 0$ (i.e. $P_f = \sigma_3$), the equation has a unique solution for $l = m = 0$ (or, alternatively, $\alpha = \beta = 90°$; see Fig. 3.7B), meaning that only discontinuities orthogonal to σ_3 can open. Finally, for $R' > 1$ (i.e. $P_f > \sigma_1$), the condition given by Eq. (5.23) is always met and any discontinuity can open independently of its orientation (also for $R' = 1$ but the case is treated thereafter).

We, therefore, focus on the cases for which $0 < R' \leq 1$ and search for the orientations of the reactivated discontinuities as a function of R' and Φ, assuming that the orientations of the principal axes of stress are known. For $R' > 0$, Eq. (5.23) can be rewritten as follows:

$$\frac{1}{R'} l^2 + \frac{\Phi}{R'} m^2 \leq 1 \qquad (5.24)$$

Eq. (5.24) represents the series of spherical caps containing the poles of the discontinuities reactivated in tension. We first note that the first-order shape of the spherical caps depends on R'. For $R' \leq \Phi$, the caps are elliptical cones accepting the minimum principal stress axis as symmetry axis, whereas for $R' > \Phi$, they are girdles about the maximum principal stress axis (Jolly and Sanderson, 1997 and Fig. 5.3). Keeping this important information in mind, we search for characterising the limits of the spherical caps, i.e. of the solution domains, setting:

$$\frac{1}{R'} l^2 + \frac{\Phi}{R'} m^2 = 1 \qquad (5.25)$$

For $0 < R' \leq \Phi$, each elliptical cone is fully defined by introducing two angles: θ_2, the angle separating the cone from σ_1 in the (σ_1, σ_3) plane; and θ_1, the angle separating the cone from σ_2 in the (σ_2, σ_3) plane (Fig. 5.3A). The respective expressions of the two angles are found by setting alternatively $l = 0$ and $m = 0$ in Eq. (5.25):

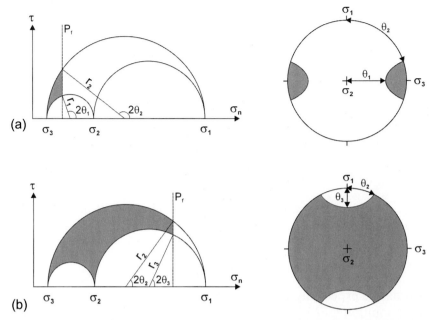

FIG. 5.3 *Grey* areas represent the poles of the planar discontinuities reactivated in tension (and in shear for most of them) for (A) $P_f < \sigma_2$ (i.e. $R' < \Phi$) and (B) $P_f > \sigma_2$ (i.e. $R' > \Phi$). Note that the spherical caps representing the collections of opened fractures take the shape of elliptical cones in the former case and of girdles in the latter case on a stereogram (Schmidt net with projection on the lower hemisphere in the present case). Each spherical cap corresponds to a specific stress state and a specific driving pressure ratio, and can be identified by two angles, either θ_1 and θ_2 or θ_2 and θ_3, depending on the value of R' with respect to Φ (see text for details).

$$\cos^2\theta_1 = \frac{R'}{\Phi} \tag{5.26}$$

$$\cos^2\theta_2 = R' \tag{5.27}$$

According to Eq. (5.26), the angle θ_1 cannot be defined if $R' > \Phi$ (i.e. if $P_f > \sigma_2$) or, conversely, when the spherical caps adopt the shapes of girdles about σ_1 (Fig. 5.3B). The treatment of these cases requires replacing θ_1 by θ_3, the angle between σ_1 and the girdle in the (σ_1, σ_2) plane, and to search for its mathematical expression.

By definition, the pole in the girdle identified by θ_3 is such that $n = 0$, $l = \cos\theta_3$ and $m = \sin\theta_3$ (i.e. $m^2 = 1 - l^2$). Thus, using Eq. (5.25), we find

$$\frac{1}{R'}l^2 + \frac{\Phi}{R'}\left(1 - l^2\right) = 1 \tag{5.28}$$

$$\frac{1}{R'}\cos^2\theta_3 + \frac{\Phi}{R'}\left(1 - \cos^2\theta_3\right) = 1 \tag{5.29}$$

and finally

$$\cos^2\theta_3 = \frac{R' - \Phi}{1 - \Phi} \tag{5.30}$$

We remark that the expression of θ_3 is consistent with the starting condition $1 \geq R' > \Phi$ (i.e. θ_3 cannot be defined for $R' < \Phi$), which, in turn, implies that the problematic case $\Phi = 1$ is safely discarded.

In summary, Eqs (5.26), (5.27) and (5.30) allow for determining the orientations of the discontinuities reactivated in tension as a function of stress anisotropy, driving pore pressure and stress orientation. The simplicity of the equations makes the formulation of the inverse problem straightforward.

Re-arranging Eqs (5.26), (5.27) and (5.30) and using double-angle formulae, we get

$$R' = \frac{1 + \cos 2\theta_2}{2} \tag{5.31}$$

$$\text{for } 0 < R' \leq \Phi \quad \Phi = \frac{1 + \cos 2\theta_2}{1 + \cos 2\theta_1} \tag{5.32}$$

$$\text{for } \Phi < R' \leq 1 \quad \Phi = 1 - \frac{1 - \cos 2\theta_2}{1 - \cos 2\theta_3} \tag{5.33}$$

These expressions show that it is theoretically possible to derive the stress state and driving pressure provided the distribution of fracture poles defines an elliptical cone or a girdle, whose characteristic angles can be accurately measured on a net.

As shown by Jolly and Sanderson (1997), the relationships devised above are easily visualised and demonstrated in the Mohr space. Consider the case $0 < R' \leq \Phi$ depicted in Fig. 5.3A. The counterpart in the Mohr space of the pole at the cone's edge, identified by θ_2 on the stereogram, is the intersection between the pore pressure line and the (σ_1, σ_3) circle (recall that this circle represents the (σ_1, σ_3) plane in the geometrical space; see Section 3.1.5 and, in particular, Fig. 3.11 for details on the construction the 3D Mohr circle). Thus, the obtuse angle between segment r_2, containing the centre of the (σ_1, σ_3) circle, and the abscissa corresponds to $2\theta_2$. Similarly, the pole identified by θ_1 on the stereogram corresponds to the intersection between the pore pressure line and the (σ_2, σ_3) circle, and the angle between segment r_1 and the abscissa is $2\theta_1$.

According to the geometrical relationships depicted in the 3D Mohr circle in Fig. 5.3A:

$$P_f - \sigma_3 = r_1(1 + \cos 2\theta_1) \tag{5.34}$$

$$\text{and} \quad P_f - \sigma_3 = r_2(1 + \cos 2\theta_2) \tag{5.35}$$

$$\text{thus} \quad \frac{r_1}{r_2} = \frac{(1 + \cos 2\theta_2)}{(1 + \cos 2\theta_1)} \tag{5.36}$$

Substituting the expression of $P_f - \sigma_3$ given by Eq. (5.35) in Eq. (5.22) and recalling that $\sigma_1 - \sigma_3 = 2r_2$, we get

$$R' = \frac{P_f - \sigma_3}{\sigma_1 - \sigma_3} = \frac{r_2(1 + \cos 2\theta_2)}{2r_2} \tag{5.37}$$

which leads to the same equation as Eq. (5.31) after simplification.

For $0 < R' \leq \Phi$, the expression of the shape ratio according to Fig. 5.3A is:

$$\Phi = \frac{\sigma_2 - \sigma_3}{\sigma_1 - \sigma_3} = \frac{r_1}{r_2} \qquad (5.38)$$

which is analogous to Eq. (5.32) after substituting the term to the right by the expression given by Eq. (5.36).

Finally, according to the geometrical relationships depicted by the 3D Mohr circle in Fig. 5.3B:

$$\sigma_1 - P_f = r_2(1 - \cos 2\theta_2) \qquad (5.39)$$

$$\text{and} \quad \sigma_1 - P_f = r_3(1 - \cos 2\theta_3) \qquad (5.40)$$

$$\text{thus} \quad \frac{r_3}{r_2} = \frac{(1 - \cos 2\theta_2)}{(1 - \cos 2\theta_3)} \qquad (5.41)$$

Noting that $r_1 = r_2 - r_3$, the expression of the shape ratio for $\Phi < R' \leq 1$ can be derived after rewriting Eq. (5.38):

$$\Phi = 1 - \frac{r_3}{r_2} \qquad (5.42)$$

We find again Eq. (5.33) after substituting the term to the right by Eq. (5.41).

In summary, it is possible, either analytically or graphically, to derive the orientations of the principal axes of stress, the shape ratio and the driving pressure ratio from a set of mode 1 or hybrid fractures. Early applications of the method were presented by Jolly and Sanderson (1997), who studied magmatic dykes, and by André et al. (2001), McKeagney et al. (2004) and Mazzarini and Isola (2007), who focussed on mineral veins.

Before discussing numerical inversion, let us examine some results from the direct problem to get a feel of the expected solutions. Fig. 5.4 shows the distributions of poles of the reactivated discontinuities as a function of R'; as expected, the greater the value of R', the wider the spherical cap containing the opened fractures, and the two extreme cases $R' = 0$ and $R' = 1$ correspond to opening limited to fractures perpendicular to σ_3 and opening to all fractures of the 3D space, respectively. As already mentioned, the shape of the spherical caps switches from conical to girdle-like for $R' > \Phi$, i.e. for $P_f > \sigma_2$. Fig. 5.5 explores the influence of Φ on the distribution of reactivated fractures. Expectedly, their poles are contained within girdles when $\Phi < R'$ and within elliptical cones otherwise. Notably, progressive increase in shape ratio, R' being constant, leads to gradual restriction of possible orientations for the reactivated fractures.

5.2 Numerical inversion

5.2.1 Bingham distribution

The only existing numerical method to reconstruct paleostresses by means of inversion of tensile fractures (Yamaji, 2016 and references therein) is based on the *Bingham*

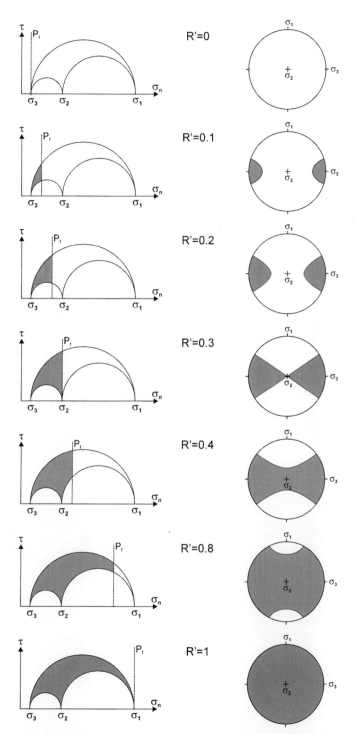

FIG. 5.4 Orientations of discontinuities reactivated in tension (and in shear for most of them) as a function of driving pressure ratio, R', and for a shape ratio value equal to $\Phi = 0.3$, depicted in grey in Mohr circles and Schmidt stereograms (projection on the lower hemisphere).

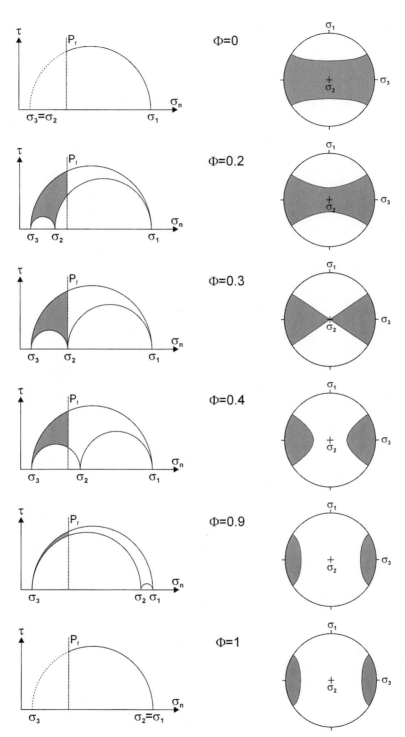

FIG. 5.5 Orientations of discontinuities reactivated in tension (and in shear for most of them) as a function of shape ratio, Φ, and for a driving pressure ratio equal to $R' = 0.3$, depicted in grey in Mohr circles and Schmidt stereograms (projection on the lower hemisphere). For Φ = 0 and Φ = 1, reactivated discontinuities are depicted by the dashed portions of the corresponding Mohr circles.

distribution (Bingham, 1974). The Bingham distribution represents an extension of the multivariate normal distribution to orientations of lines distributed in the 3D space, which for convenience are projected on a Schmidt net (e.g. Cheeney, 1983, p. 119–127). In Earth sciences, the Bingham distribution was originally employed to characterise dispersion of paleomagnetic vectors. It is a mathematical tool to quantify the symmetry of a set of axial data. More specifically, it permits to characterise any orthorhombic distribution of points on a sphere (i.e. any distribution accepting three mutually orthogonal principal planes of symmetry). The corresponding poles may form elliptical clusters, girdles or, as end-member case, be uniformly distributed (Fig. 5.6A).

Given a set of axial data, the aim of this method is to determine (1) the intersections between the principal planes of symmetry, i.e. the three principal axes of symmetry; and (2) three algebraic quantities, i.e. the concentrations κ_1, κ_2 and κ_3, honouring the distribution of the data. The three principal axes of symmetry, t_1, t_2 and t_3 point respectively to directions of minimum, intermediate and maximum concentrations of poles on a net

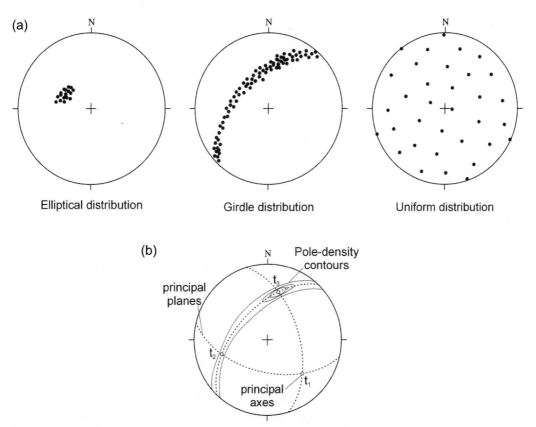

FIG. 5.6 (A) Synthetic examples of pole distributions, the first two following orthorhombic symmetries; and (B) application of Bingham statistics to determine the three principal planes and the three principal axes of symmetry. The principal axes t_1, t_2 and t_3 point to directions of minimum, intermediate and maximum concentrations of poles, respectively. Projection is on the lower hemisphere of a Schmidt stereogram.

(Fig. 5.6B). The three concentrations are measures of data dispersion parallel to the three principal planes, i.e. κ_1, κ_2 and κ_3 furnish measures of data dispersion from t_1 to t_3, from t_2 to t_3 and from t_1 to t_2, respectively.

Here, we examine only relevant mathematical aspects of the Bingham distribution; further details can be found in Love (2007). The general form of the probability density function of the Bingham distribution is

$$P_B(\mathbf{n}|K, E) = \frac{1}{A} \exp\left(\mathbf{n}^T E^T K E \mathbf{n}\right) \tag{5.43}$$

where $\mathbf{n}$ is an orientation vector, A is a normalisation constant, E is an orthogonal matrix admitting the principal axes of symmetry as eigenvectors (i.e. furnishing the attitude of the Bingham distribution), ^{T} indicates matrix transpose and K is the concentration matrix:

$$K = \begin{pmatrix} \kappa_1 & 0 & 0 \\ 0 & \kappa_2 & 0 \\ 0 & 0 & \kappa_3 \end{pmatrix} \tag{5.44}$$

Eq. (5.43) shows, as expected, that the Bingham distribution is antipodally symmetric, i.e. $P_B(\mathbf{n}|K,E) = P_B(-\mathbf{n}|K,E)$. It can also be shown that the distribution is invariant with respect to the sum of the concentrations. Therefore, the concentrations can be conveniently ordered:

$$\kappa_1 \leq \kappa_2 \leq \kappa_3 = 0 \tag{5.45}$$

The latter choice permits to characterise the symmetry of any set of axial data with only five parameters, the orientations of the three principal axes and the two concentrations κ_1 and κ_2. Assuming that the orientations of the principal axes of symmetry are known, the values of κ_1 and κ_2 determine the shape of the distribution as visualised on a Schmidt stereogram (Fig. 5.7). If $\kappa_1 = \kappa_2 = \kappa_3 = 0$, the distribution is uniform. Elliptical distributions of poles correspond to $\kappa_1 \leq \kappa_2 < 0$, with the particular case of $\kappa_1 = \kappa_2 < 0$ indicating circular distributions; and girdle distributions require $\kappa_1 \ll \kappa_2 < 0$ and low absolute values for κ_2 (Yamaji and Sato, 2011).

5.2.2 Formulation of the inversion problem

a) Justification for the Bingham distribution

Distributions of poles of planar discontinuities reactivated in tension (Fig. 5.4) and the Bingham distribution (Fig. 5.7) show a clear analogy, suggesting that Bingham statistics is an appropriate mathematical tool to quantify the reduced stress tensor and the driving pressure ratio related to a given set of tensile fractures. The analogy is better perceived when assimilating the poles of reactivated fractures to normal stress vectors for which $\sigma_n \leq P_f$. For instance, if $P_f \geq \sigma_1$, i.e. the condition $\sigma_n \leq P_f$ is always met, the distribution is uniform (Fig. 5.4) and can potentially be described by a Bingham distribution with $\kappa_1 = \kappa_2 = 0$ (Fig. 5.7). If P_f is slightly higher than the minimum principal stress, fracture poles make a tight cluster around the orientation of σ_3; this configuration resembles a

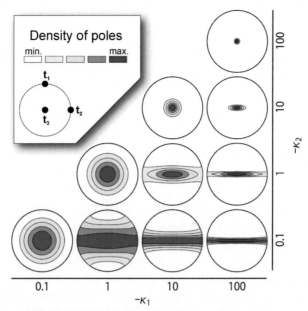

FIG. 5.7 Shape of (Bingham) pole distributions as a function of concentrations κ_1 and κ_2 as visualised on Schmidt stereograms. Note that $\kappa_1 \leq \kappa_2 < 0$ characterises elliptical distributions of poles and equal concentrations indicate circular distributions, girdle distributions being described by $\kappa_1 \ll \kappa_2 < 0$ and low absolute values for κ_2. The orientations of the three principal axes of symmetry are depicted in the inset. *Modified from Yamaji, A., Sato, K., 2011. Clustering of fracture orientations using a mixed Bingham distribution and its application to paleostress analysis from dike or vein orientations. J. Struct. Geol. 33, 1148–1157.*

Bingham distribution with $\kappa_1 \approx \kappa_2 \ll 0$. Intermediate distributions adopt the shapes of either elliptical clusters or girdles with obvious orthorhombic symmetries, which once again can be described by the Bingham distribution.

To demonstrate the appropriateness of the Bingham distribution, we follow the steps taken by Yamaji et al. (2010). For the sake of the demonstration, let us recall the expression of the reduced stress tensor (Eq. (4.5) in Section 4.1.3):

$$T_r = \begin{pmatrix} 1 & 0 & 0 \\ 0 & \Phi & 0 \\ 0 & 0 & 0 \end{pmatrix}$$

As we already know (Chapter 4), the expression of the stress tensor in the physical coordinate system is given by

$$T_{0r} = E^T T_r E \tag{5.46}$$

where E is the rotation matrix defining the change of coordinate system and the uppercase symbol ^{T} indicates matrix transpose.

Eq. (3.53) defining the magnitude of the normal stress on a plane can also be written in matrix form:

$$\sigma_n = \mathbf{n}^T \boldsymbol{\sigma} \qquad (5.47)$$

where $\mathbf{n}$ and $\boldsymbol{\sigma}$ are column vectors, and by definition

$$\boldsymbol{\sigma} = T_{0r} \, \mathbf{n} \qquad (5.48)$$

Combining Eqs (5.46)–(5.48), we get

$$\sigma_n = \mathbf{n}^T E^T T_r E \mathbf{n} \qquad (5.49)$$

We have considered until now that opening of a set of tensile fractures occurs in response to a single pulse of pore pressure, P_f, being equal to the normal stress acting on the planes of the least optimally orientated discontinuities (i.e. the ones whose poles define the limits of the elliptical clusters and girdles in Figs 5.4 and 5.5). This assumption is implicit in the theory by Jolly and Sanderson (1997) and should a priori lead to uniform distributions of poles within the elliptical clusters and girdles (Fig. 5.8).

As pointed out by Yamaji et al. (2010), optimally orientated planar discontinuities are more likely to be reactivated than the ones submitted to higher σ_n values during a single pressure pulse. This statement appears physically sound because slight variations in attitude along non-optimally orientated discontinuities are expected to result in local normal stress values higher than the pore pressure and, therefore, in local barriers, impeding the fluids to propagate further to open these discontinuities. Thus, in the case of a single pressure pulse, reactivation probability is reasonably expressed by some decreasing function of σ_n.

Furthermore, geological observations show that the opening of a set of tensile (and/or hybrid) fractures is driven by a series of pore pressure pulses taking place during the tectonic event corresponding to the fracture set (e.g. Ramsay, 1980; Bahat and Engelder, 1984; Sibson, 1990; Cox et al., 1991). The identification of distinct mineral growth events in veins (Ramsay, 1980) or successive plumose structures and arrest marks along a single joint surface (Bahat and Engelder, 1984) are convincing pieces of evidence for syn-tectonic pore fluid fluctuations. It is reasonable to assume that moderate pressure pulses are more frequent than high-pressure pulses. Therefore, the distribution of reactivated discontinuities is expected to be inversely proportional to σ_n in the case of various pressure pulses.

In brief, the probability distribution of planar discontinuities reactivated in tension can be written as a decreasing function of σ_n, or as a decreasing function of the expression of σ_n as given in Eq. (5.49):

$$P(\mathbf{n}) \propto f\left(\mathbf{n}^T E^T T_r E \mathbf{n}\right) \qquad (5.50)$$

The function describes the expected decrease in the concentration of poles of reactivated fractures from the orientation of σ_3 to that of σ_1. However, the exact form of this function is not known. Sato et al. (2013) explored different forms of decreasing functions and showed that, in general, they led to similar results and, in particular, to very close

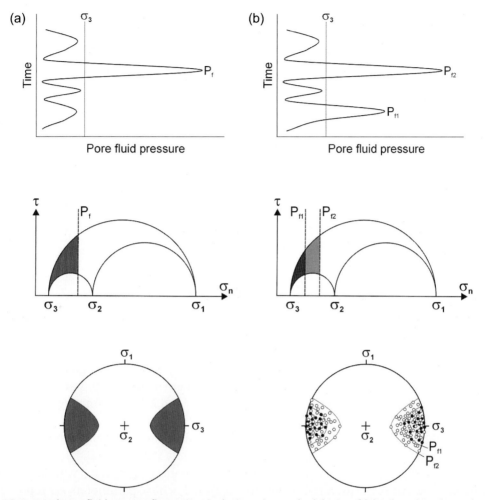

FIG. 5.8 Assumed pore fluid pressure fluctuations in the inversion methodologies of (A) Jolly and Sanderson (1997) and (B) Yamaji et al. (2010). Jolly and Sanderson's method considers a single pressure pulse, $P_f \geq \sigma_3$, resulting in uniform distribution of fracture poles within an elliptical cluster or girdle (in *dark grey* in (A)). Yamaji et al. assume successive pulses of overpressure with different magnitudes during the tectonic event. The lowest pressure, P_{f1}, has the potential to reactivate only a restricted population of fractures (*black poles* in (B)), whereas the highest pressure, P_{f2}, has the potential to reactivate a wider range of fractures (*white poles* in (B)). Thus, the density of fracture poles is expected to be higher within the elliptical cluster or girdle corresponding to P_{f1}. For the sake of clarity, only two pressure pulses are depicted; the combination of a relatively high number of pulses leads to a gradual decrease in pole density away from σ_3.

orientations for the three principal axes of stress. We follow Yamaji et al. (2010) and assume a simple exponentially decreasing function:

$$P(\mathbf{n}) \propto \exp\left(-\mathbf{n}^T E^T T_r E \mathbf{n}/\Lambda\right) \qquad (5.51)$$

where Λ is a constant indicating the relative number of high pore pressure fluid pulses with respect to low ones.

Comparing Eqs (5.51) and (5.43) demonstrates that $P(\mathbf{n})$ corresponds to the Bingham distribution $P_B(\mathbf{n}\,|\,T_r, E)$, whose symmetry axes are the three principal axes of stress, i.e. $t_1 = \sigma_1$, $t_2 = \sigma_2$ and $t_3 = \sigma_3$ and whose concentration matrix is given by

$$K = -\frac{1}{\Lambda}T_r \tag{5.52}$$

Replacing K and T_r by their respective matrix forms, their corresponding third rows and columns being neglected, we find

$$\begin{pmatrix} \kappa_1 & 0 \\ 0 & \kappa_2 \end{pmatrix} = -\frac{1}{\Lambda}\begin{pmatrix} 1 & 0 \\ 0 & \Phi \end{pmatrix} \tag{5.53}$$

$$\text{Thus,}\quad \kappa_1 = -\frac{1}{\Lambda} \tag{5.54}$$

$$\text{and}\quad \kappa_2 = -\frac{\Phi}{\Lambda} \tag{5.55}$$

$$\text{and finally}\quad \Phi = \frac{\kappa_2}{\kappa_1} \tag{5.56}$$

b) Numerical inversion

According to previous mathematical developments, the orientations of the principal axes of stress and the shape ratio can be derived from the Bingham distribution characterising the distribution of poles of discontinuities reactivated in tension. The goal of the numerical inversion is therefore to find the best-fitting Bingham distribution. To this end, Yamaji et al. (2010) proposed to optimise the two matrices E and K by maximising a logarithmic likelihood function.

A *likelihood function* (or *likelihood*) is a function designed to estimate the goodness of fit of a statistical model to a sample of data (Fisher, 1921, 1922). In practice, the likelihood function is written as a function of the observed N variables, $\mathbf{x}_1, \mathbf{x}_2 \ldots \mathbf{x}_N$, with the assumption that they satisfy the probability density function P, whose parameters are represented by θ:

$$L = \prod_{i=1}^{N} P(\mathbf{x}_i|\theta) \tag{5.57}$$

The optimal values of the parameters of the probability density function (usually indicated by circumflexes) are found for the maximum value taken by L:

$$L_{\max} = \prod_{i=1}^{N} P(\mathbf{x}_i|\hat{\theta}) \tag{5.58}$$

The reader will find further details on the likelihood in classical textbooks about statistics (e.g. van den Bos, 2007, pp. 100–120). For the sake of clarification, an illustrative example of application of the likelihood theory is given in Fig. 5.9.

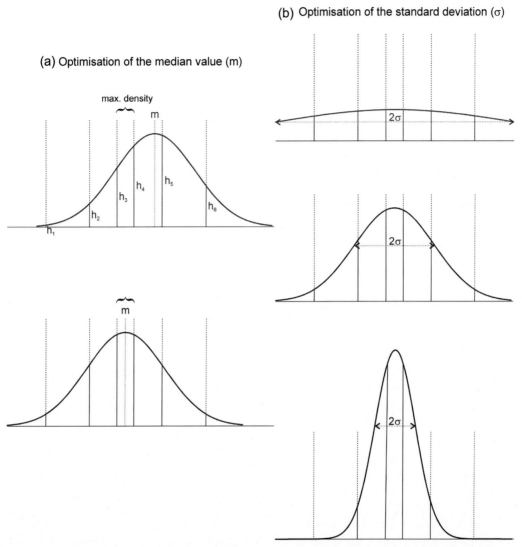

(a) Optimisation of the median value (m)

(b) Optimisation of the standard deviation (σ)

FIG. 5.9 Example of likelihood function to estimate the fit of a statistical model to a distribution of six variables (dotted black bars). In the present case, the likelihood function to be maximised is $L = \prod_{i=1}^{6} h_i$, and the statistical model is a Gaussian function defined by two parameters, its median (*m*) and its standard deviation (*σ*). The optimal values of these two parameters are determined by maximising *L*. **(A)** Optimisation of *m*: *L* is maximised when a maximum of variables adopt relatively high values, i.e. when the Gaussian curve is adjusted such that *m* corresponds to the centre of the maximum density of the distribution (compare the *red* and *blue curves*). **(B)** Optimisation of *σ*: However, the previous operation does not constrain the spread of the Gaussian curve (i.e. *σ*). If *σ* is relatively high (*red curve*), the six variables adopt close but relatively low values and their sum, *L*, is low. If *σ* is relatively low (*black curve*), the variables close to *m* are maximised but at the expense of the other variables, especially the most distal ones that can be dramatically reduced; therefore, *L* is not maximised. The Gaussian curve that maximises *L* and corresponds to the optimal *σ* value (*blue curve*) has to be found between these two end-member cases.

The likelihood being always positive and the natural logarithm being a monotonically increasing function, it is often more convenient to maximise the corresponding *logarithmic likelihood function*:

$$\mathscr{L} = \ln(L) = \sum_{i=1}^{N} \ln[P(\mathbf{x_i}|\theta)] \tag{5.59}$$

Replacing P, $\mathbf{x_i}$ and θ respectively by P_B, $\mathbf{n_i}$, and K and E in Eq. (5.59) leads to the expression of the logarithmic likelihood function we need to maximise to optimise E and K, the parameters of the Bingham distribution of the poles of the N measured tensile fractures:

$$\mathscr{L}(KE) = \sum_{i=1}^{N} \ln[P_B(\mathbf{n_i}|KE)] \tag{5.60}$$

Combining Eqs (5.43) and (5.60), we get

$$\mathscr{L}(KE) = \sum_{i=1}^{N} \mathbf{n_i}^T E^T KE\mathbf{n_i} - N \ln A \tag{5.61}$$

The constant A is a function of the concentrations of poles and is calculated by integrating the probability density over the unit sphere. The start of the maximisation of the logarithmic likelihood function requires the setting up of initial values for the concentrations $\kappa_1 = \kappa_2 = -10$. As the second step, the matrix E is initiated using the symmetry axes of the orientation matrix of the fracture poles.

When the optimisation of E and K is completed and the Bingham distribution of the measured tensile fractures is fully determined, the orientations of the three principal axes of stress are directly derived and the shape ratio is calculated using Eq. (5.56). The maximum (normalised) normal stress found for the set of analysed fractures is taken equal to the (normalised) value of the maximum pore fluid pressure and is used to calculate the driving pressure, R' (Eq. (5.22)).

c) Error estimation

As already pointed out, the Bingham distribution is characterised by five independent parameters whose values are given by the two matrices K and E. To estimate the robustness of the paleostress inversion results, it is particularly convenient to handle these five parameters as the five components of a unit vector, $\mathbf{u_j}$, in the 5D space (see complete mathematical formulation in Sato and Yamaji, 2006b; Yamaji and Sato, 2011). Thus, each vector $\mathbf{u_j}$ represents a specific reduced stress tensor.

The error analysis consists of evaluating the relative stability of the three principal axes of stress and evaluating the shape ratio around their determined position or, conversely, to search for the 95% confidence intervals of each component of the reduced stress tensor. To these aims, the *bootstrap method* (Efron, 1979) is employed. In brief, the bootstrap method is a statistical method based on random sampling with replacement of a population and subsequent generation of synthetic (i.e. bootstrap) samples. For instance, consider the population (1, 2, 3, 4, 5); after re-sampling and replacement, we obtain the bootstrap samples (3, 2, 5, 4, 5) and (4, 4, 2, 4, 1). Repeating the operation many times, we get a significant number of bootstrap samples that, in turn, can be used to infer statistical parameters such as median, confidence level or variance.

The population to be analysed here is the set of the N fractures considered for the paleostress inversion. First, the N data are re-sampled N_b times with $N_b \gg N$. A collection of N_b bootstrap samples is thus obtained. N_b stress states are subsequently computed and their corresponding N_b $\mathbf{u_j}$ vectors derived.

The optimal reduced stress tensor is represented by the mean 5D vector:

$$\overline{\mathbf{u}} = \frac{\sum_{j=1}^{N_b} \mathbf{u_j}}{\left| \sum_{j=1}^{N_b} \mathbf{u_j} \right|} \tag{5.62}$$

The collection of $\mathbf{u_j}$ stands for the dispersion of the results around the optimal solution $\overline{\mathbf{u}}$ or, in other words, for the degree of uncertainty of the paleostress inversion. Let us calculate the angular distance separating each vector from the mean vector $\overline{\mathbf{u}}$:

$$\Theta_j = \cos^{-1}\left(\mathbf{u_j}^T \overline{\mathbf{u}}\right) \tag{5.63}$$

Θ_j varies from $0°$, when the two vectors coincide (i.e. perfect fit), to $180°$, if they point in opposite directions (i.e. complete misfit) in the 5D space. The 95% confidence region is constructed by excluding the 5% $\mathbf{u_j}$ vectors presenting the largest misfits, and corresponding stress tensors are calculated for the 95% remaining vectors. Finally, the 95% confidence regions of the three principal axes of stress and of the shape ratio are derived from the optimal solution and the 95% best-fitting tensors.

However, it is more convenient to use a single scalar value to access the robustness of the paleostress inversion. The mean angular stress difference, originally introduced for fault slip inversion techniques by Yamaji et al. (2006), is additionally employed:

$$\overline{\Theta} = \frac{\sum_{j=1}^{N_b} \Theta_j}{N_b} \tag{5.64}$$

where $\overline{\Theta} = 90°$ corresponds to a random distribution of the stress states (i.e. fully unconstrained results), $\overline{\Theta} = 0°$ indicates that all stress states are identical (i.e. fully constrained results); thus, the quality of the results is high for low values of $\overline{\Theta}$.

d) Extension to heterogeneous data

Commonly, a collection of fractures measured in the field contains subsets belonging to distinct stress events even when the minerals hosted by the fractures are identical. The mixing of these data subsets leads to meaningless 'average' stress states. To separate heterogeneous datasets during inversion, Yamaji and Sato (2011) extended the numerical approach introduced by Yamaji et al. (2010).

As previously mentioned, the Bingham distribution is characterised by five independent parameters whose values are given by the two matrices K and E. It is mathematically more convenient to handle these five parameters as the five components of a position vector, $\mathbf{v}$, in the 5D space. Therefore, the Bingham distribution is rewritten as

$$P_B(\mathbf{n}|K, E) = P_B(\mathbf{n}|\mathbf{v}) \tag{5.65}$$

Let us consider a collection of tensile fractures involving M distinct subsets. Each fracture subset forms an elliptical cluster or a girdle corresponding to a specific Bingham distribution (i.e. to specific values of the five parameters of the Bingham distribution). The superposition of all the elliptical clusters/girdles results in a cloud of poles, which by assumption is characterised by the *mixed Bingham distribution*, the weighted sum of the Bingham distribution corresponding to each data subset:

$$P_{mB}(\mathbf{n}|\theta, \varpi) = \sum_{j=1}^{M} \varpi_j P_B(\mathbf{n}|\mathbf{v_j}) \tag{5.66}$$

where ϖ_j is the compounding ratio of the mixing coefficient of the jth Bingham distribution, whose five parameters are given by the vector $\mathbf{v_j}$; ϖ and θ represent all compounding ratios and the whole collection of $\mathbf{v_j}$, respectively.

By definition, each ϖ_j reflects the relative contribution of its corresponding subset of poles to the mixed Bingham distribution such that

$$0 < \varpi_j \leq 1 \tag{5.67}$$

$$\text{and} \quad \sum_{j=1}^{M} \varpi_j = 1 \tag{5.68}$$

Similarly to the treatment of a single elliptical cluster or girdle (Eq. (5.60)), we maximise the logarithmic likelihood function of the mixed Bingham distribution:

$$\mathcal{L}(\theta\varpi) = \sum_{i=1}^{N} \ln\left[P_{mB}(\mathbf{n_i}|\theta\varpi)\right] \tag{5.69}$$

where N is the number of fractures, and ϖ and θ are the parameters to be optimised.

In the present case, the maximisation involves determining the total number of Bingham distributions, defining the mixed Bingham distribution, and quantifying each of them, i.e. separation of the dataset into distinct clusters (or clustering, see Section 4.2.6). Equations such as (5.69) are traditionally maximised using an *Expectation–Maximisation (EM) algorithm* (Dempster et al., 1977). An EM algorithm is an iterative method involving an E-step, where an expected logarithmic likelihood function is calculated based on a set of parameters, and an M-step, where the parameters are optimised through maximisation of the expected logarithmic likelihood function. The following E-step is then performed with the optimised parameters.

Yamaji and Sato (2011) implemented the EM method to maximise Eq. (5.69) as described here. First, θ is initialised by generating randomly M $\mathbf{v_j}$ vectors around the origin of the 5D parameter space. This step is equivalent to set an a priori number of Bingham distributions (i.e. elliptical clusters and/or girdles) and to give a priori values to the five parameters (i.e. to define a priori E and K matrices) quantifying each of them. The first E-step evaluates the membership of each datum (i.e. each one of the N fracture poles) according to

$$m_{i,j} = \frac{P_B(\mathbf{n_i}|\mathbf{v_j})}{\sum\limits_{j=1}^{M} P_B(\mathbf{n_i}|\mathbf{v_j})} \tag{5.70}$$

$$\text{with } 0 \leq m_{i,j} \leq 1 \tag{5.71}$$

$$\text{and } \sum_{j=1}^{M} m_{i,j} = 1 \tag{5.72}$$

Eq. (5.70) indicates the degree of membership of the ith datum to the jth elliptical cluster or girdle, whose Bingham distribution is defined by $\mathbf{v_j}$. For example, let us assume that the whole dataset involves two elliptical clusters or girdles (i.e. $M=2$) and $m_{i,1}=0.3$. The latter value indicates that the ith fracture pole in the dataset belongs to both the first and second group of poles with probabilities of 30% and 70%, respectively. Once the membership is evaluated for all the fracture poles of the dataset, the set of mixing coefficients, ϖ, is derived and, finally, the expected logarithmic likelihood function is calculated.

The M-step consists of optimising θ while keeping ϖ unchanged by maximising the logarithmic likelihood function previously determined in the course of the E-step. To these aims, Yamaji and Sato (2011) employed the simplex method from Nelder and Mead (1965), but other maximisation strategies can be adopted as well. Once θ is optimised the parameter is updated and a new E-step is run. The EM algorithm is terminated when the variations in θ and ϖ become insignificant after two successive iterations.

However, EM algorithms converge in most cases to local maxima depending on the initial parameter values (i.e. the random set of $\mathbf{v_j}$ here). The search for the global maximum of the logarithmic likelihood function requires relatively high numbers of EM runs with different initial parameter values. Yamaji (2016) proposed, however, an improvement to the traditional EM scheme.

Increasing M in Eq. (5.66), i.e. the number of Bingham distributions, improves automatically the goodness of the mixed Bingham distribution (i.e. the more elliptical clusters and girdles are considered, the better the clustering of the data). The latter operation is equivalent to increasing the number of independent variables to fit a model. To remediate this potential pitfall, a *Bayesian information criterion* (Schwarz, 1978) is traditionally used:

$$\text{BIC} = -2\mathscr{L}(\hat{\theta}\hat{\varpi}) + (6M - 1)\ln N \tag{5.73}$$

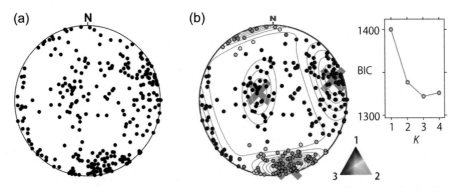

FIG. 5.10 (A) Collection of 286 tensile fractures measured by Shanley and Mahtab (1976); the distribution of poles points unambiguously to heterogeneous data (i.e. multiple stress events). (B) inversion of the data results in three clusters corresponding to three distinct stress states as indicated by the minimum value of the BIC; for simplification, only the orientations of the axes of minimum principal stress (i.e. grey crosses) are depicted. Colours indicate memberships of fracture poles and contours confidence intervals. *Modified after Yamaji, A., Sato, K., 2011. Clustering of fracture orientations using a mixed Bingham distribution and its application to paleostress analysis from dike or vein orientations. J. Struct. Geol. 33, 1148–1157.*

The BIC is written as a function of the logarithmic likelihood and degree of freedom of the mixed Bingham distribution (i.e. the term 6M-1). The last term of Eq. (5.73) acts as a penalty against increasing degree of freedom or, conversely, the number of elliptical clusters and girdles. In practice, the solution corresponding to the minimum BIC is preferred (Fig. 5.10).

5.3 Further remarks

Paleostress reconstruction based on tensile fractures is a recent class of methods compared to the more classical fault slip inversion methods. Relatively few case studies have been published until now and, to the best knowledge of the author, no systematic testing and benchmarking has been conducted yet. The physical background of these methods has been rarely discussed or refined since the publication of the seminal paper of Jolly and Sanderson (1997)—the assumption of multiple overpressure events introduced by Yamaji et al. (2010) being the only exception (Fig. 5.8). In brief, inversion of tensile fractures is still a tool under development (at the date of publication of this book), and its basic principles and inherent limitations deserve discussion.

At first view, it is less fieldwork-intensive to measure attitudes of tensile fractures than fault slip data. This pleasant advantage is, however, counterbalanced by the relatively large amounts of data needed to constrain the shapes of the elliptical cones or girdles to be analysed. One may conceive that paleostress determinations on only few tens of measurements are not robust unless the cluster of poles is particularly tight. The robustness of the analysis deteriorates further in case of distributions of poles presenting relatively large spreads (e.g. girdles). We face here the very first limitation of the approach: robust determination of the shape of the distributions of poles of tensile fractures (and ultimately of the parameters of the reduced stress tensor and of the driving pressure) demand high amounts of data, typically various tens to hundreds of measurements.

The previously mentioned issue can be easily solved when field conditions are favourable. More problematic is, however, the limitations imposed by the theoretical background of the method as it is the case for any mathematical method and, in particular, the assumption of pre-existing planar discontinuities with evenly distributed attitudes. This condition enforcing rock isotropy as a prerequisite to the application of the method, whereas anisotropy (i.e. well-defined preferential orientations of discontinuities) being definitively more common in nature, needs to be regarded with much circumspection.

The characterisation of the shapes of the elliptical cones and girdles devised in Fig. 5.4 depends on the variety of orientations of the available discontinuities. Restriction of orientations for discontinuities susceptible to open under the prescribed stress and pore pressure conditions implies a reduction in the size of the cluster of poles and, in general, modification of its geometry. These alterations may in turn result in significant errors in the computation of the driving pressure ratio and of the stress parameters, as suggested in Fig. 5.11. In summary, the safe application of paleostress inversion of tensile fractures

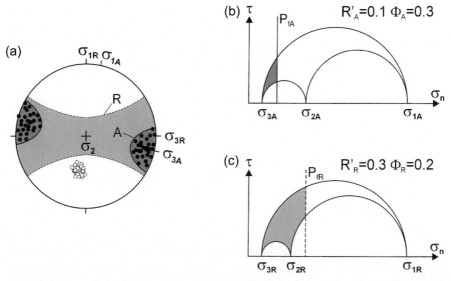

FIG. 5.11 Hypothetic example showing how non-uniform distributions of pre-existing planar discontinuities in the rock may affect paleostress determinations. (A) Representation of the problem on a Schmidt stereogram (lower hemisphere projection) and (B) biased and (C) accurate results in the Mohr space. The rock mass contains two populations of discontinuities depicted by their respective clusters of poles (i.e. ~NNE–SSW sub-vertical and ~ESE–WNW north dipping discontinuities). *Black and white dots* represent the poles of reactivated and non-reactivated discontinuities, respectively. The *dark grey* elliptical cone is defined based on the distribution of the reactivated discontinuities; as described in this chapter, the resulting paleostress and driving pressure ratio are indicated by the suffix 'A' (i.e. 'Apparent'). The *light grey* girdle shows the ideal pole distribution corresponding to the actual paleostress and pore pressure conditions (indicated by the suffix 'R' for 'Real'), i.e. the distribution that would have been found if the starting assumption of uniform distribution of discontinuities in the 3D space would have been met. Comparison of the 'Apparent' and 'Real' solutions suggests that rock anisotropy, a common situation in nature, may seriously deteriorate not only the accuracy of the driving pressure ratio but also the goodness of the paleostress determination. To note, only the orientation of σ_2 is accurately resolved in the present case.

seems to be limited to seldom occurrences of rocks containing uniformly distributed discontinuities only.

The discussion appears to take a rather pessimistic shade. Shall we abandon the method? To the opinion of the author, constructive rethinking of its conceptual framework may root the approach on firmer ground. Instead of assuming the opening of pre-existing discontinuities, it is proposed that the tensile fractures involved in the analysis are neoformed.

The latter assumption justifies the inescapable condition for isotropy, pre-existing discontinuities being replaced by Griffith-type micro-cracks (see Section 3.2.3). Admittedly, it is expected that some micro-cracks are more susceptible to grow according to their respective orientations (i.e. lengths), depending in turn on the loading history of the rock. However, their propagation paths tend to adapt progressively to the ambient stress conditions; thus, they can be viewed as microscopic seeds of fractures, themselves able to develop in all directions allowed by the ambient stress and pore pressure conditions. Reformulated in these terms, the background assumption ensures that a collection of fractures measured in the field define properly the shapes of the elliptical cones or girdles involved in the analysis and, therefore, reflect the values of the five parameters, in particular, those of the reduced stress tensor, with satisfying confidence.

As usual, the gain is not warranted without some sacrifice. Replacing Jolly and Sanderson's assumption by the previous one implies that pore pressure acts not only against normal stress on fracture walls, which was a pleasant feature of the theory, but also against the tensile strength of the rock to promote opening. Clearly, it means that condition (5.1) should be changed to

$$P_f \geq \sigma_n + T \tag{5.74}$$

where T is uniaxial tensile strength.

As a consequence, the driving pressure ratio needs to be re-defined:

$$R'' = \frac{(P_f - T) - \sigma_3}{\sigma_1 - \sigma_3} \tag{5.75}$$

Therefore, the driving pressure ratio as derived from the inversion of tensile fractures does not reflect a normalised relative value of pore pressure but a normalised relative value of the difference between pore pressure and tensile strength. In other words, the ratio reflects a minimum estimate for pore pressure, which needs to be corrected for the tensile strength of the rock.

Further reading

Baer, G., Beyth, M., Reches, Z., 1994. Dikes emplaced into fractured basement, Timna Igneous Complex, Israel. J. Geophys. Res. Solid Earth 99, 24039–24050.

Delaney, P.T., Pollard, D.D., Ziony, J.I., McKee, E.H., 1986. Field relations between dikes and joints: emplacement processes and paleostress analysis. J. Geophys. Res. Solid Earth 91, 4920–4938.

Jolly, R.J.H., Sanderson, D.J., 1997. A Mohr circle construction for the opening of a pre-existing fracture. J. Struct. Geol. 19, 887–892.

Yamaji, A., 2016. Genetic algorithm for fitting a mixed Bingham distribution to 3D orientations: a tool for the statistical and paleostress analyses of fracture orientations. Island Arc 25, 72–83.

Yamaji, A., Sato, K., 2011. Clustering of fracture orientations using a mixed Bingham distribution and its application to paleostress analysis from dike or vein orientations. J. Struct. Geol. 33, 1148–1157.

Yamaji, A., Sato, K., Tonai, S., 2010. Stochastic modeling for the stress inversion of vein orientations: paleostress analysis of Pliocene epithermal veins in southwestern Kyushu, Japan. J. Struct. Geol. 32, 1137–1146.

Inversion of calcite twins

6.1 Calcite twins: Definitions and general aspects

Under certain conditions, mineral grains can exhibit internal changes in lattice orientation. These changes occur between *host* and *twinned crystals*, themselves separated by common crystallographic planes (*composition* or *twin planes*). Twins may result from crystallisation (i.e. growth twinning) or from deformation of a pre-existing grain (Fig. 6.1). The latter twinning process is referred to as *mechanical* or *deformation twinning*. Relatively complex twinning implies rotation of the twinned crystal with respect to the host according to a symmetry axis perpendicular to the twin plane.

In its simplest form, mechanical twinning involves bending of the mineral in conditions approximating simple shear, i.e. the lattice of the twinned crystal is rotated with respect to that of the host according to an angle characteristic of the mineral (Fig. 6.2). The rotation itself is the result of translation of the lattice parallel to one favourably oriented crystallographic plane (i.e. one of the *twin planes* of the mineral) and according to a specific direction (i.e. the *twinning direction* associated with the specific twin plane). This deformation mechanism is referred to as *twin gliding* and leads to a final configuration where the twin crystal is the mirror image of the host crystal with respect to the twin plane. No slip of the twin on the host occurs. In addition, the mechanism of twinning does not involve friction or dilation (thus implying that twinning is insensitive to confining pressure) nor activation energies (thus implying that twinning does not depend on temperature theoretically, though a weak dependence has been found in laboratory experiments; see below) and, finally, twinning does not depend on strain rate. Mechanical twinning of crystals appears to be primarily governed by differential stress.

Calcite is a mineral particularly prone to mechanical twinning and, as such, a wealth of research has been conducted on calcite twinning since the late 19th century. Mechanical twinning of calcite (i.e. *e-twinning*) occurs parallel to three specific crystallographic twin planes (i.e. *e-planes*). In agreement with the third-order symmetry of calcite, the three e-planes are arranged symmetrically around the optical axis, *C*, each of the three normals of the e-planes leaning ∼26° away from *C* (Fig. 6.3). The three *twinning directions* strictly follow the lines of intersection between *e*-planes and *r*-cleavage planes, or conversely, the orthogonal projections of *C* on the three *e*-planes, and e-twinning is only permitted along these three twinning directions. When *C* is rotated to vertical, the senses of motion on the *e*-planes resemble that of reverse dip-slip faults, the upper parts of the twinned crystals moving upwards, towards the *C* optical axis. Practically, one calcite twin is identified and its orientation is determined by measuring the angle between the pole of the e-plane and the *C*-axis of the twin using an optical microscope with a three-axis U-stage.

FIG. 6.1 Microphotograph of mechanical twins affecting calcite; note multiple sets. *Courtesy of Olivier Lacombe, Sorbonne Université, Paris.*

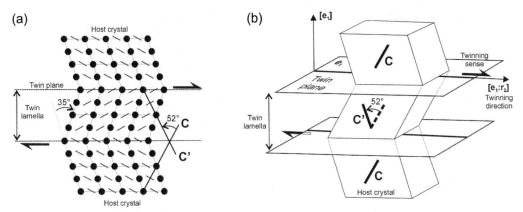

FIG. 6.2 Mechanical twinning of crystals (calcite in the present case) represented (A) in 2D and depicting lattice deformation (the short lines are parallel to the CO_3^{2-} complexes) and (B) in 3D at crystal scale. Mechanical twinning results in bending of the crystal's lattice (up to 35° in the present case) under conditions approximating simple shear parallel to the twin plane, e.g. e_1, and according to a specific twinning direction, i.e. $[e_1:r_2]$. The deformation implies rotation of the optical axis of the twinned crystal (*C'*) with respect to the optical axis of the host crystal (*C*), which is 52° in the case of calcite. $[e_1]$ is the normal of the twin plane e_1.

E-twinning is the dominant crystal-plastic deformation for calcite below ~400°C; above this temperature, other mechanisms become progressively more efficient to accommodate deformation and eventually take over. As such, e-twins are the most abundant microstructures in carbonate rocks in the upper brittle crust. Though calcite twinning is practically temperature-insensitive, the morphology of twin lamellae varies with temperature and can be used as a temperature proxy: thin twins form at relatively low

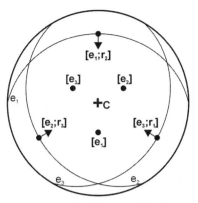

FIG. 6.3 Stereographic projection depicting the specific symmetry of the three *e*-planes (or twin planes) of calcite (i.e. e_1, e_2 and e_3) with respect to the optical axis, C. The poles of the planes are disposed symmetrically 26° from C. Note that when C is rotated to vertical, the sense of motion according to the *e*-planes resembles that of reverse dip-slip faults, as suggested by the inwards arrows. [e_1:r_2], [e_2:r_3] and [e_3:r_1] represent the three allowed twinning directions. Projection on the lower hemisphere of a Schmidt net.

temperatures, whereas thick and curved ones are diagnostic of higher temperatures (Burkhard, 1993). Shear strain related to e-twinning is low and bound to 0.35 for each lamella (Fig. 6.2); further strain needs to be accommodated by other mechanisms (e.g. fracturing of the grain).

As already pointed out, mechanical twinning depends primarily on applied differential stresses. In the case of e-twinning, experimental laboratory work has shown that twin crystals become more abundant in a deformed sample with increasing differential stresses. Moreover, twin gliding on e-planes seems to be initiated when reaching a given threshold of ~10 MPa, which corresponds to the *critical resolved shear stress* (*CRSS*) on the *e*-planes ('resolved' stands for 'according to twinning directions'; Fig. 6.4).

It is worth noting that lab experiments have demonstrated that twin abundance is also proportional to grain size—relatively large grain sizes favouring e-twinning—and strongly suggest that the CRSS is not constant but inversely proportional to grain size. The assumption of a universal CRSS is to be treated with much caution and, in particular, rocks composed of relatively large grains of calcite with homogeneous sizes are to be preferred for paleostress inversion of calcite twins. Finally, the CRSS appears to be weakly dependent on temperature, the higher the temperature the lower the CRSS, and on strain as a consequence of strain hardening.

6.2 Calcite twins as paleopiezometers: Early studies

Early e-twinning paleopiezometry methods are fundamentally based on the relationship between abundance of twins and the magnitude of applied differential stresses on the rock sample as established in the lab. The method pioneered by Jamison and Spang

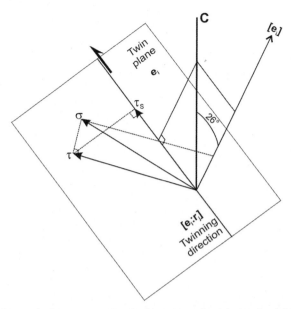

FIG. 6.4 Stress condition for e-twinning to occur. E-twinning occurs when the resolved shear stress, τ_s, i.e. the shear stress component along the twinning direction, exceeds the critical resolved shear stress (CRSS). In general, τ_s does not correspond to the maximum shear stress, τ. The latter configuration is met only if the stress vector, σ, is in the plane containing both the optical axis (C) and the normal of the e-plane ([e_i], where i = 1, 2 or 3 in the case of calcite).

(1976) consists of counting the relative percentages of activated e-planes. Its main assumptions are (1) a constant CRSS, (2) stress uniformity in the aggregate and at grain scale and (3) evenly distributed calcite grains (thus evenly distributed e-planes). In more detail, resolved shear stresses on e-planes increase with increasing applied differential stresses (i.e. $\sigma_1 - \sigma_3$), the CRSS being first reached on optimally oriented e-planes. When differential stress magnitude is further augmented, the CRSS is also reached on less favourably oriented e-planes, which are subsequently activated. Thus, the relative number of twins on 0, 1, 2 or 3 e-planes is a function of applied differential stress magnitude, which can therefore be estimated after counting the twins in the studied sample. However, by considering implicitly that all twins in the sample belong to a single tectonic event (i.e. single stress event characterised by a uniaxial stress state), the method suffers from a severe limitation in the common case of polyphase deformation.

Other approaches, such as that introduced by Rowe and Rutter (1990), focus on statistical twinning features such as twinning incidence (i.e. relative number of grains of a given size showing twins), twin volume fraction (relative twinned volume of a crystal) and twin density (number of twins per mm), which were experimentally found to be proportional to applied differential stresses. Nevertheless, further experimental work demonstrated that twinning incidence and twin volume fraction were very sensitive to grain size. Twin density appears to be a more reliable stress gauge but overestimates differential stress magnitudes for temperatures below $\sim$200°C (e.g. Lacombe, 2010).

6.3 Inversion methods

6.3.1 Turner's method

The very first stress inversion method using calcite twins was pioneered by Turner (1953) and is restricted to the estimation of the orientations of the principal stress axes and, to some extent, the degree of stress anisotropy. Turner's method assumes that, during deformation, 'compression' and 'tension' axes are optimally oriented to maximise shear along the twinning direction of a given twinned e-plane, or conversely, 'compression' and 'tension' axes project on the twinning direction and are at 45° with respect to the twinned e-plane, as depicted in Fig. 6.5. This optimal configuration implies specific orientations between the optical C-axis of the host crystal and the 'compression' and 'tension' axes, the latter making acute angles of 71° and 19°, respectively, with the optical C-axis (Fig. 6.5A). Hence, measurement of the respective orientations of the pole of the twinned e-plane and the optical C-axis of the host crystal allows for determining the orientations of the 'compression' and 'tension' axes (Fig. 6.5B). The rotation of the optical axis consecutive to twinning (Fig. 6.2) is such that the optical C'-axis of the twinned crystal is flipped towards the direction of the 'compression' axis with respect to the C-axis.

Strictly speaking, the so-called 'compression' and 'tension' axes are symmetry elements of the system, similar to the P and T axes in the right dihedra method (see Section 4.3), and correspond to stress or strain axes only in the particular case of perfect alignment. Consequently, one has to be much aware that the symmetry pattern, potentially reconstructed from the distribution of 'compression' and 'tension' axes associated with a collection of

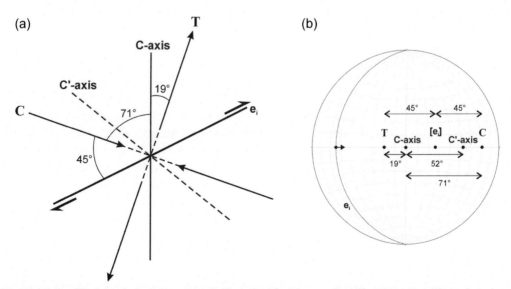

FIG. 6.5 (A) Simplified sketch depicting optimal orientations for 'compression' (*C*) and 'tension' (*T*) axes in order to promote e-twinning; (B) projection of the elements shown in (A) on the lower hemisphere of a Schmidt net. C-axis and C'-axis are the optical axes of the host and the twin, respectively; e_i and $[e_i]$ represent the e-plane and its pole, respectively.

twinned e-planes, reflects both the initial symmetry of the lattice and the symmetry of the applied stress, both influences being a priori of equal importance.

Turner's method furnishes direct estimates of stress orientations only if the lattice presents a relatively simple initial symmetry (and provided that this initial symmetry can be inferred); the stress analysis of more complex symmetries requires examination of various symmetry elements of the system and of carefully selected grains (see details at pp. 413–421 of Turner and Weiss, 1963). A random distribution of e-planes in the studied sample represents, for example, a simple case where the pattern of the lattice is spherical and, hence, any other pattern of symmetry as evidenced by the distribution of 'compression' and 'tension' axes reflects the one of the applied stress. In such a case, the maxima in the distributions of 'compression' and 'tension' axes, as visualised on a net, correspond to the orientations of σ_1 and σ_3, respectively.

The result is obviously a direct consequence of the existence of a CRSS. The e-planes that are preferentially twinned are the ones that are optimally oriented (i.e. their related 'compression' and 'tension' axes are parallel to σ_1 and σ_3, respectively) and exert a heavy weight on the focussing of the maxima in the statistical distributions of 'compression' and 'tension' axes, whereas e-planes with less favourable orientations contribute less in focussing the distributions and, finally, e-planes with unfavourable orientations do not contribute at all. Assuming a constant CRSS, the precision of the determination is dependent on the magnitude of the differential stress applied to the rock. For instance, if the CRSS is equal to 10 MPa and $\sigma_1 - \sigma_3 = 20$ MPa, the resolved shear stress reaches its allowed maximum value of 10 MPa only for optimally oriented e-planes. This results in very tight maxima for the orientations of the 'compression' and 'tension' axes on the net. As differential stress increases, the CRSS is progressively reached on less optimally oriented e-planes and, consequently, the obtained maxima spread and become more diffuse.

In summary, Turner's method is an elegant and easy-to-implement method and provides reasonable first-order results. As stated above, the method becomes more difficult to use when the initial symmetry pattern of the lattice is relatively complex and differs considerably from that of the applied stress. Superposition of various stress states on the studied sample results in a similar limitation, where different symmetry systems become difficult to separate. The optimal configuration between 'compression' and 'tension' axes and activated e-planes is reminiscent of that between P and T seismological axes and nodal planes as restored from P-wave first arrivals (see Section 4.3), allowing for the use of the more advanced right dihedra method to paleostress inversion of calcite twins (Pfiffner and Burkhard, 1987).

6.3.2 The calcite stress inversion technique

a) Background principles

The so-called "CSIT" (i.e. Calcite Stress Inversion Technique, term coined by O. Lacombe and co-workers) is the most rigorous, robust and widely used calcite twin inversion method. It was progressively developed by Laurent et al. in the 80s (Laurent, 1984; Laurent et al., 1981, 1990; Tourneret and Laurent, 1990), based on the paleostress inversion scheme originally introduced by Etchecopar (1984). The background assumptions of the CSIT are (1) stress uniformity at the grain scale and (2) unique CRSS (i.e. τ_a) for the studied sample. Considering both twinned and untwinned e-planes, the principle of the CSIT is to find a unique stress tensor (T) such that the calculated maximum resolved shear stresses (τ) respect the following set of conditions:

$$\tau < \tau_a \text{ for the untwinned planes and } \tau \geq \tau_a \text{ for the twinned planes} \tag{6.1}$$

As already mentioned, twinning is independent of pressure; in other words, an infinite number of stress tensors sharing the same deviatoric component but with differing isotropic ones fit the relationships (6.1). The implication of the latter physical condition is that, theoretically, only the deviatoric component of T (i.e. T_D) can be restored by means of inversion of calcite twins. A constant τ_a could be assumed (and was introduced in pioneer versions of the CSIT, e.g. Laurent et al., 1990) as an additional constraint to the inversion problem, allowing for the calculation of five out of six parameters of the stress tensor. However, the result becomes strongly submitted to some assumed 'universal' CRSS, whereas laboratory experiments suggest that the latter value is variable depending on grain size, strain and, to some extent, temperature. Practically, in the 80s, the choice led to a dramatic lengthening of the computation time and was found inappropriate in the case of polyphase deformation (at least because the CRSS is expected to increase with strain).

To overcome these limitations, a better strategy consists of adapting fault slip inversion methods to the problem of stress determination from calcite twins, taking advantage of the similarities existing between the two deformation mechanisms (i.e. considering twinning directions on activated e-planes equivalent to striae on fault planes). The CIST involves the principles of the fault slip inversion method originally introduced by Etchecopar et al. (1981) and Etchecopar (1984).

Similar to most classical fault slip inversion methods (see Chapter 4), Etchecopar's method restores the four parameters of the reduced stress tensor (i.e. three angles accounting for the orientations of the three principal axes of stress and the shape ratio, Φ, representing stress anisotropy):

$$T_r = \begin{pmatrix} 1 & 0 & 0 \\ 0 & \Phi & 0 \\ 0 & 0 & 0 \end{pmatrix} \tag{6.2}$$

$$\text{with } 0 \leq \Phi \leq 1 \tag{6.3}$$

It is worth noting that the form of the reduced tensor given in Eq. (6.2) implies that the reduced differential stress (i.e. the difference between the maximum and minimum principal stresses; see Chapter 3) is scaled and equal to 1. Therefore, the resolved shear stress on any plane is (see Section 4.1.3)

$$-0.5 \leq \tau_S \leq 0.5 \tag{6.4}$$

We have already emphasised that inversion of twin data is bound to deliver the deviatoric component of the stress tensor only. Hence, a wise way to handle the numerical problem consists of considering only the reduced deviatoric component in the equations to be solved. The latter component is simply calculated by subtracting isotropic stress (i.e. the trace of T_r divided by 3; see Chapter 3, Eq. 3.33) from T_r:

$$T_{Dr} = T_r - \frac{1+\Phi}{3} I \tag{6.5}$$

where I is the unit matrix.

As such, the form of the reduced deviator is

$$T_{Dr} = \begin{pmatrix} 2-\Phi & 0 & 0 \\ 0 & 2\Phi-1 & 0 \\ 0 & 0 & -(\Phi+1) \end{pmatrix} \tag{6.6}$$

Practically, the CSIT solves the four parameters of the deviator given in Eq. (6.6) by means of inversion of twin data. We should note the two following points. (1) The condition (6.4) remains untouched because, as seen in, e.g. Section 4.1.3, shear stress is insensitive to the addition or subtraction of isotropic stresses. (2) The use of the reduced deviator introduces automatically an additional constraint to the problem; the trace of this tensor must be equal to 0 (i.e. by definition isotropic stresses are null for a deviator).

As discussed previously, assuming an a priori value for the CRSS is problematic both from computational and conceptual aspects. On the contrary, the physical concept of CRSS can be used to better constrain the inversion problem. Once the inversion has been performed, the four parameters of T_{Dr} determined and a subsequent collection of τ_S calculated for the twinned e-planes, an astute and sensible assumption can be introduced: the smallest positive value found for τ_S on the twinned planes (i.e. τ_{Sa}) represents a scaled value of the CRSS.

Thus

$$\frac{\sigma_1 - \sigma_3}{\tau_a} = \frac{(\Delta\sigma)\text{scaled}}{\tau_{Sa}} = \frac{1}{\tau_{Sa}} \tag{6.7}$$

and

$$\sigma_1 - \sigma_3 = \frac{\tau_a}{\tau_{Sa}} \tag{6.8}$$

In brief, if a value for the CRSS (i.e. τ_a) is assumed after inversion of the twin data, it is possible to estimate a fifth parameter of the deviatoric stress tensor, the differential stress. Rigorously speaking, the CSIT determines by means of numerical inversion four parameters of the stress tensor, which explains why computation is faster with respect to previous methods, but derives a fifth one from the results of the inversion.

b) The penalisation function
Condition (6.1) is, however, too stringent for the scheme detailed above to work on natural cases. Because of measurement errors, polyphase deformation and/or local stress deviations, it is very unlikely that, in a natural rock sample, a perfect solution (i.e. a unique

tensor predicting resolved shear stresses in agreement with all the twinned and untwinned e-planes) emerges from a series of twinned and untwinned e-planes.

As a best case, we would like to find and retain a solution, for which departures with respect to the ideal perfect solution are minimal and, to these aims, to introduce some quantitative measure of how many e-planes are accurately described by the solution and of how much the identified faulty planes diverge from this ideal solution. One might spontaneously think that the more untwinned e-planes are predicted to be twinned and the more twinned planes are predicted untwinned, the less reliable is the restored stress tensor. Therefore, any quantitative measure for the reliability of the result, for example a *penalisation function*, should a priori involve both twinned and untwinned e-planes.

However, in the case of polyphase deformation, a penalisation function accounting for by the twinned planes is misleading. For instance, if the studied sample contains two sets of twins corresponding to two tectonic events, the inversion might restore accurately the stress tensor related to one of the two events and naturally predict as untwinned (most of) the twinned e-planes associated with the other event. As a consequence, the penalisation function will suggest a poor result, whereas the stress tensor is accurately determined for one of the two sets of twins (provided that no or only proportionally few untwinned e-planes are themselves predicted to be twinned).

To overcome the latter limitation, Tourneret and Laurent (1990) focused exclusively on untwinned e-planes and introduced the following penalisation function:

$$F = \sum_{j=1}^{n} (\tau_{Sj} - \tau_{Sa}) \tag{6.9}$$

$$\text{with } \tau_{Sj} > \tau_{Sa} \tag{6.10}$$

where n represents the number of untwinned *e*-planes, which are predicted to be twinned, and τ_{Sj} the resolved shear stresses computed for these specific *e*-planes.

The penalisation function, F, equals 0 for the ideal perfect case (i.e. none of the untwinned planes is predicted to be twinned; Fig. 6.6) and takes a positive value otherwise. The higher the value taken by F, the less reliable is the result of the inversion; hence, the inversion procedure consists of finding the stress tensor corresponding to a minimum value of F. The choice made by Tourneret and Laurent (1990) of involving exclusively the untwinned *e*-planes in the minimisation process is particularly elegant in the case of polyphase samples, because the constraint that the untwinned *e*-planes should not be twinned applies to all the successive stress events that produced the twins measured in the sample.

c) The inversion problem

Based on the principles and the penalisation function described above, the inversion procedure aims at sorting out a maximum of twinned and untwinned *e*-planes such that their corresponding resolved shear stress magnitudes, τ_S, are respectively equal or larger and lower than the CRSS, τ_{Sa} (Fig. 6.6).

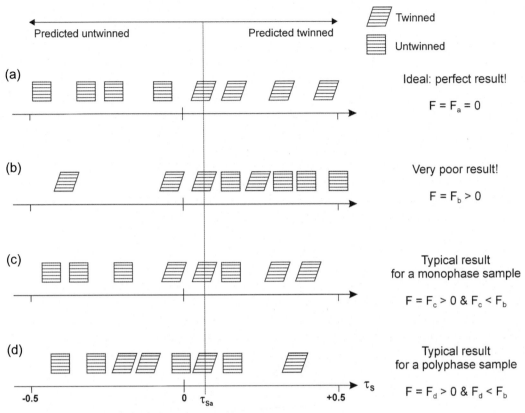

FIG. 6.6 Schematic representation of the background principles of the penalisation function *F* introduced by Tourneret and Laurent (1990). (A) Ideal result: all twinned and all untwinned planes fall respectively to the right (i.e. predicted twinned) and to the left (i.e. predicted untwinned) of the scaled value of the CRSS, τ_{Sa}. (B) Very poor result (i.e. high value for the penalisation function): 100% of untwinned planes fall to the right. (C) Typical (good) result for a monophase sample: 75% of the twinned planes fall to the right and 75% of the untwinned ones to the left. (D) Typical (good) result for a polyphase sample: 50% of the twinned planes fall to the right and 75% of the untwinned ones to the left, the 50% of the twinned planes falling to the left representing a distinct phase of deformation. Note that cases (A) and (B) are unlikely and shown only for clarification. A unique τ_{Sa} has been assumed for the sake of simplification. τ_S represents scaled resolved shear stress.

As the first step, forward computations are performed. This step allows for a substantial reduction in the computing time. A large number (e.g. 100–1000) of randomly selected stress tensors is applied to the dataset of twinned and untwinned e-planes, resolved shear stresses are computed for each *e*-plane and each stress tensor and the minimisation function, *F,* is subsequently determined for each tested stress tensor. The stress tensor corresponding to the minimum value of *F* or, conversely, to the best sorting of the e-planes, is then automatically selected. The second step consists of the inversion sensu stricto. The selected stress tensor is optimised by local exploration around minimum values of *F*; the computations deliver the final result when the minimum *F* value is found.

Nevertheless, the procedure described above does not allow for rigorous treatment of polyphase samples and, if applied to the whole collection of *e*-planes in such a sample, would lead to the calculation of a meaningless 'bulk' stress tensor. To tackle this critical problem, Tourneret and Laurent (1990) adapted the mathematical method of Etchecopar et al. (1981), originally introduced to discriminate stress tensors reconstructed from poly-phase fault slip measurements. The background principle of the separation stands once again on a pragmatic application of condition (6.1). The latter condition must be satisfied for all (as much as possible) untwinned e-planes but only for part of the twinned ones. Prac-tically, the problem consists of determining the relative proportion, P_S, of twinned e-planes, which are compatible with the stress tensor. A starting value for P_S is set for the computa-tions and progressively modified while tracking the evolution of different parameters (i.e. trends and plunges of the maximum principal axes, shape ratio and τ_{Sa}), including the pena-lisation function. The result is obtained when a majority of the twinned e-planes are accounted for by the stress tensor while the value of F remains low. The twinned *e*-planes belonging to this first solution are removed from the dataset and a second tensor is searched for with the remaining twinned and all the untwinned *e*-planes as new input. The procedure is repeated until the relative proportion of twinned planes becomes insignificant.

Calcite inversion methods have practically remained unchanged for years until the recent renewal of interest in the topic. Recent improvements of the CSIT focused mainly on the crit-ical problem of separating multiple paleostresses in samples containing polyphase calcite twin data (Parlangeau et al., 2018). Of particular significance for the theoretical validation of the CSIT, Yamaji (2015) explored systematically the sensitivity of the results in the 5D hyperspace of the five reconstructed parameters (i.e. orientations of the three principal axes of stress, shape ratio and differential stress). The study demonstrated mathematically the robustness of the CSIT in determining stress orientations. It also pointed out the sensitivity of the restored differential stress to both the actual differential stress level, at which the e-twins were formed (the higher the level, the poorer the result), and sampling bias.

6.4 Validation of the calcite stress inversion technique

Numerous applications of the CSIT and similar calcite stress inversion methods are described in modern literature (see also Section 9.3); most of the studies furnished geo-logically sound results, adding support to the validity of these methods. In particular, the work of Lacombe et al. (1989, 1990) was amongst the very first ones to address the validity of the CSIT by means of systematic comparison with the results obtained independently using paleostress inversion of fault slip data.

Lacombe et al. (1990) focussed on six sites located in the Burgundy Platform, France (Fig. 6.7). The selection of the study area was motivated by the presence of a weakly deformed Middle-to-Upper Jurassic carbonate platform, with mostly horizontal layers (save at the southernmost site, i.e. Taxenne, where the sediments are involved in the fold-and-thrust Jura system), particularly suitable for both fault slip and calcite twin ana-lyses. The occurrence of numerous quarries operated in the Jurassic limestones

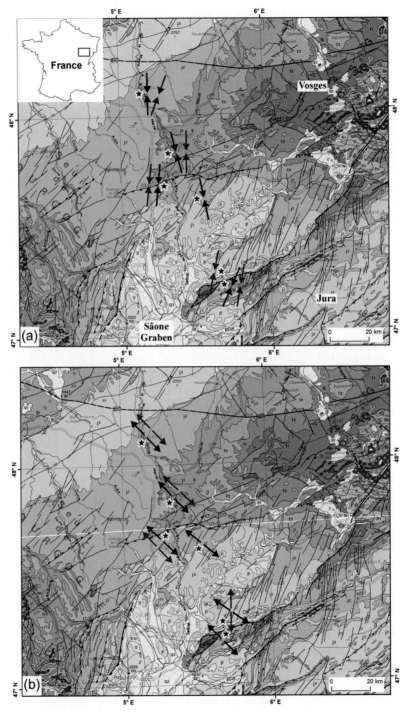

FIG. 6.7 Stress directions obtained by means of inversion of fault slips (*black*) and calcite twins (*red*). (A) N—S late Eocene (Pyrenean) compression. (B) NW—SE Oligocene extension.

(Continued)

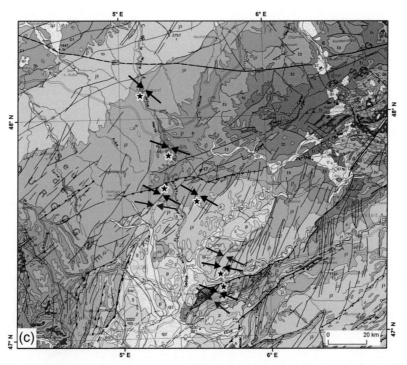

FIG. 6.7, CONT'D   (C) WNW–ESE Miocene compression. The study was conducted at six sites located in Burgundy (France): Taxenne, Montagney, Champlitte, Prauthoy, St Geosmes and Chaumont (from south to north). Original data from Lacombe et al. (1990), geological map after BRGM (http://geoservices.brgm.fr/geologie, SCAN_F_GEOL1M layer).

represented an additional advantage. The studied area was located at the north-eastern tip of the Oligocene Sâone Graben, a member of the extensive Western European Rift system (see also Section 9.1) and NW of the Miocene Jura fold-and-thrust belt. It was thus anticipated that the study would resolve and identify the different tectonic phases and associated paleostress states pertaining to the post-Jurassic regional evolution. The studied area was located on the transition zone between the Sâone Graben and the Upper Rhine Graben, whose western flank corresponds to the Vosges Mountains.

The well-established fault slip direct inversion method from Angelier (Angelier, 1990 and Section 4.2.3) was used to benchmark the results of the CSIT. Systematic fault slip measurements and measurements of stylolithes and veins (where available) were conducted in parallel to sampling for calcite twin analyses.

The analyses evidenced in particular three major tectonic phases, which were sorted out according to relative chronology criteria gathered in the field: (1) N—S compression, (2) NW–SE extension and (3) WNW–ESE compression. The three tectonic phases were well resolved both by the fault slip inversion and the CSIT (Fig. 6.7), albeit not systematically at each studied site. The N—S compression was mainly characterised by a strike-slip stress regime and was found consistent with abundant N—S stylolithic peaks measured during the course of the fieldwork. Based on correlations with regional structural data, the N—S compression was proposed to correspond to the late Eocene Pyreneo–Provençal phase

Table 6.1   Directions of the stress axes (σ_1 for the two compressive phases and σ_3 for the extensional one) determined at each studied site and according to the two methods employed.

Azim. σ_1/σ_3 in °	N–S comp.			NW–SE ext.			WNW–ESE comp.		
	FS	CT	Diff.	FS	CT	Diff.	FS	CT	Diff.
1. Taxenne	205	188	17	146			108	113	5
2. Montagney		186		134	182	48		103	
3. Champlitte	174			130			110		
4. Prauthoy	186	185	1	131			113	93	20
5. St Geosmes	175	179	4	136	135	1	105		
6. Chaumont		200		132	130	2			
Average	185	188	7	135	149	17	109	103	13

FS, fault slip inversion; CT, calcite twin inversion.

Table 6.2   Shape ratios (Φ) determined at each studied site and according to the two methods employed.

Φ	N–S comp.			NW–SE ext.			WNW–ESE comp.		
	FS	CT	Diff.	FS	CT	Diff.	FS	CT	Diff.
1. Taxenne	0.1	0.1	0.00	0.3			0.3	0.6	0.30
2. Montagney		0.6		0.3	0.6	0.30		0.3	
3. Champlitte	0.5			0.3			0.4		
4. Prauthoy	0.4	0.02	0.38	0.3			0.5	0.5	0.00
5. St Geosmes	0.5	0.4	0.10	0.2	0.5	0.30	0.4		
6. Chaumont		0.7		0.2	0.4	0.20			
Average	0.38	0.36	0.16	0.27	0.50	0.27	0.40	0.47	0.15

FS, fault slip inversion; CT, calcite twin inversion.

(i.e. collision of the Iberian microplate with Eurasia). The NW–SE extension was in agreement with the observation of numerous NE–SW veins. Lacombe et al. (1990) interpreted the latter extension direction as a local deviation, along the transition zone between the Sâone and Upper Rhine grabens, of the well-documented E–W Oligocene extension, which was responsible for the formation of both grabens. According to relative chronology data, the WNW–ESE compression (strike-slip regime) was the youngest event. It was correlated with the Miocene thrusting and folding of the Jura Mountains.

Admittedly, the number of determinations was relatively modest, but the agreement between average stress directions for each tectonic phase, either σ_1 for the two phases of compression or σ_3 for the phase of extension as derived from both methods, was nevertheless remarkable (Table 6.1). A detailed look at the results, site by site and specifically at those sites where stress tensors could be determined by the two methods, confirmed the goodness of the result: for seven out of eight cases, departures in stress directions were found to be less than 20°, and the results were in excellent agreement (i.e. differed by less than 5°) for five cases.

Furthermore, it is interesting to compare the shape ratios, Φ, derived from both methods (Table 6.2). At first glance and on average, the ratios appear to be in good agreement, at least for the compressive phases. However, after careful examination of the results, it becomes clear that the agreement is only apparent; for each specific site (and each tectonic phase), the values differ by more than 20% (i.e. five out of eight cases) in general.

The poor agreement in shape ratios between both methods suggests at first that one or both has limited power to resolve Φ. In general, data dispersion has a strong influence on the determination of the latter value. Data dispersion arises from imprecise measurements and, more importantly, from erroneous inclusion of measurements related to distinct deformation phases into the dataset. Both fault slip and calcite twin inversion methods are sensitive to the latter perturbation. The automatic selection of twin data, which is implemented in the CSIT, makes the latter method particularly prone to data contamination, and we may reasonably suspect that the shape ratios are normally less well resolved with respect to fault slip inversion methods. Nevertheless, one should also keep in mind that stress magnitudes do not stay constant throughout the duration of a given tectonic event, primarily because of exhumation/burial or repetitive cycles of elastic stress loading and stress drop consecutive to yielding. As a consequence, the shape ratio does not remain constant but fluctuates with time. Finally, the shape ratio as determined by inversion methods represents stress anisotropy at yield (i.e. when faulting or twinning occurs) and not the average stress anisotropy. We may thus anticipate that even for ideal datasets, the two methods result in different shape ratios that, in turn, reflect different deformation mechanisms.

In conclusion, Lacombe et al. (1990) demonstrated that inversion of calcite twins, using the CSIT, resolves accurately major tectonic phases of regional significance in weakly deformed terranes and, particularly, is able to discriminate and separate successive stress tensors. However, the CSIT is unable to furnish relative chronologies between the determined stress tensors; the latter information must be gathered through complementary studies in the field.

Further reading

Burkhard, M., 1993. Calcite twins, their geometry, appearance and significance as stress-strain markers and indicators of tectonic regime: a review. J. Struct. Geol. 15, 351–368. The Geometry of Naturally Deformed Rocks.

Jamison, W.R., Spang, J.H., 1976. Use of calcite twin lamellae to infer differential stress. GSA Bull. 87, 868–872.

Lacombe, O., 2010. Calcite twins, a tool for tectonic studies in thrust belts and stable orogenic forelands. Oil Gas Sci. Technol. – Revue d'IFP Energies Nouvelles 65, 809–838.

Rowe, K.J., Rutter, E.H., 1990. Palaeostress estimation using calcite twinning: experimental calibration and application to nature. J. Struct. Geol. 12, 1–17.

Tourneret, C., Laurent, P., 1990. Paleo-stress orientations from calcite twins in the North Pyrenean foreland, determined by the Etchecopar inverse method. Tectonophysics 180, 287–302.

Turner, F.J., 1953. Nature and dynamic interpretation of deformation lamellae in calcite of three marbles. Am. J. Sci. 251, 276–298.

Turner, F.J., Weiss, L.E., 1963. Structural Analysis of Metamorphic Tectonites. McGraw-Hill, New York.

Determining paleostresses with 'incomplete' data

7.1 Forward modelling

7.1.1 Fault slip modelling

Forward modelling of fault slips is based on the classical equation of Bott (1959) relating the maximum resolved shear stress on a fault plane to the four parameters of the reduced stress tensor (i.e. orientations of the three principal axes of stress and shape ratio; see Eq. 4.13). This type of method represents the simplest, not to say the most primitive, one amongst the collection of existing methods to deal with stress–slip relationships (e.g. Pascal and Angelier, 2003).

The method is weak when attempting to determine the values of the four stress parameters from fault slip data, as numerous calculation runs are needed and, in the end, the result is a rough approximation of the solution, whose accuracy depends chiefly on the tenacity of the user. Nevertheless, forward modelling becomes a more interesting technique when the goal of the study is the estimation of (paleo)slip directions on fault surfaces that are not directly accessible to the geologist, e.g. the ones routinely imaged in the subsurface by means of geophysical surveys. In particular, strike-slip components of relatively modest magnitudes are very difficult to detect using geophysical methods. Fault slip modelling can shed light on the potential existence of strike-slip movements and furnish estimates on their magnitudes. Additional advantages of forward modelling methods are short computation times, easy incorporation of sensitivity analyses in the calculations and straightforward interpretation of the results (Pascal, 2004).

Fig. 7.1 depicts fault slip forward modelling of the major fault segments of the Frøy Field, which is located on the eastern flank of the Viking Graben in the northern North Sea. The goal of the study was to estimate potential lateral offsets of oil reservoirs (Pascal et al., 2002). The map shows the simplified structural configuration of the top Brent (Middle Jurassic); fault attitudes and their normal components, as well as their Late Jurassic age, were determined from 3D seismic surveys.

To compensate for the relative lack of quantitative information, this type of study has to rely heavily on the expertise of the geologist and the extent of regional geological knowledge at hand. The Viking Graben is an aborted Late Jurassic continental rift trending mainly NNE–SSW. The latter structural trend differs markedly from the structural grain of the underlying basement (i.e. N–S, ENE–WSW and minor WNW–ESE), suggesting that the NNE–SSW faults are products of the rifting itself. Therefore, the observations point to a

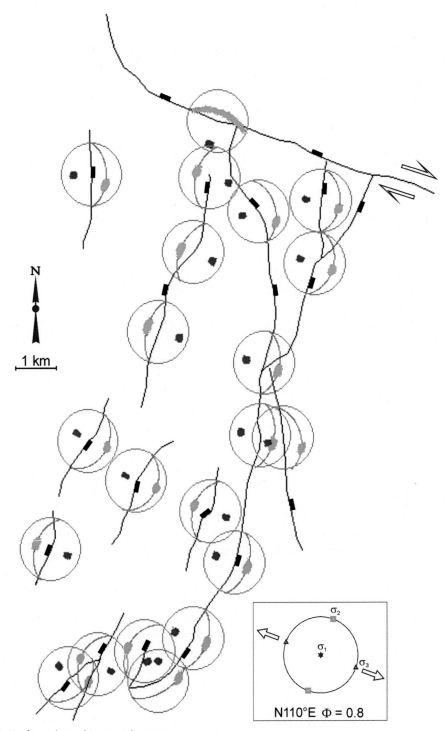

FIG. 7.1 See figure legend on opposite page

normal stress regime with σ_3 oriented ~WNW–ESE at the time of rifting in the Late Jurassic. Assuming Andersonian conditions, the latter inference allows for estimating the most probable orientations for the three principal axes of paleostress (Fig. 7.1). It thus becomes straightforward to evaluate (1) the sensitivity of the computed fault slip vectors with respect to uncertainties on stress orientation, fault dip and fault strike and (2) the influence of the shape ratio, and ultimately, derive reasonable estimates of potential strike-slip components.

The method can also be used to constrain the four parameters of the reduced stress tensor in favourable case studies. As for fault slip inversion methods, the presence of distinct fault orientations in the dataset permits reducing the collection of compatible stress tensors. In the example given in Fig. 7.1, one fault strikes WNW–ESE, which is nearly perpendicularly to the average strike of most of the faults of the Frøy Field. Moreover, the structural analysis of the field (Pascal et al., 2002) suggests a considerable dextral strike-slip along this fault during the rift event. This qualitative observation adds further constraints on the orientation of σ_3: the trend of the minimum principal axis of paleostress could not have exceeded N112°E (i.e. the average strike of the normal-dextral fault); otherwise, the fault would have shown a sinistral strike-slip component. Similar lines of thinking lead to the conclusion that σ_3 had to remain close to parallel to the fault and, finally, that the shape ratio had to be relatively high in order to promote significant strike-slip (Fig. 7.2). In brief, the previous qualitative arguments limit reasonably well the range of the most likely paleostress tensors at the location of the Frøy Field during Late Jurassic rifting.

7.1.2 Slip tendency analysis

a) Principles of fault slip tendency analysis

Slip tendency analysis (Morris et al., 1996) is a refinement of fault slip forward modelling in which friction is accounted for. The goal of the approach is to identify the planar discontinuities that are most susceptible to be reactivated as faults when submitted to a given stress. The method is mostly used to predict present-day fault stability and seismicity (e.g. Worum et al., 2004) and is often applied in the context of the oil industry or geothermal operations (e.g. Martínez-Garzón et al., 2016), though it may also be employed to identify what fault orientations had been preferentially reactivated in response to certain

---

FIG.7.1 Forward modelling of Late Jurassic slip vectors along the major faults of the Frøy Field, Viking Graben, northern North Sea. The traces of the normal faults were imaged by seismic reflection surveys at depths corresponding to the top of the Middle Jurassic Brent Formation (i.e. ~3 km below sea bed) and strikes and dips were estimated from the 3D geological model of the oil field (see Pascal et al., 2002 for details); solid rectangles indicate hanging walls. Fault slips were computed according to the stress state depicted in the inset and assuming errors of ± 0.1 for the shape ratio, Φ, and ± 10 degree for the trend of σ_3 and the dips and strikes of the modelled fault segments. The clouds of *red poles* represent the poles of the fault planes and their related orientation uncertainty. The clouds of *green poles* represent collections of possible slip vectors. Slip is poorly resolved for the WNW–ESE normal-dextral fault, which is nearly parallel to σ_3; however, only about half of the determinations are in agreement with the inferred dextral strike-slip motion. Projection is on the lower hemisphere of a Schmidt net.

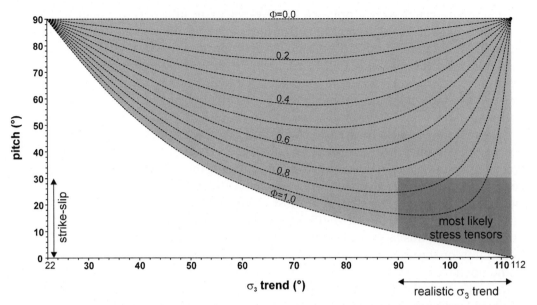

FIG. 7.2 Forward modelling of the variation in slip orientation along the WNW–ESE normal-dextral fault depicted in Fig. 7.1 as a function of Φ and σ_3, assuming Andersonian normal stress regime (see text for details). The analysis restricts the collection of stress tensors compatible with considerable dextral strike-slip along the fault (*dark grey box*). The *open circle* on the abscissa indicates that the function is discontinuous for σ_3 parallel to the fault (i.e. for N112°E in the present case); for this case, the slip jumps to pure dip-slip (*solid circle*).

paleostress regimes. However, the method requires knowledge of stress magnitudes as a prerequisite.

Slip tendency of a planar discontinuity was defined by Morris et al. (1996) as the ratio between shear and normal stress:

$$T_S = \frac{\tau}{\sigma_n} \tag{7.1}$$

where T_s is positive, considering that $\sigma_n > 0$, in the case of shear reactivation of a discontinuity, and that the sign of τ is meaningless and conveniently set positive, as long as we are solely concerned with the reactivation criterion (see e.g. the discussion on Mohr circles in Section 3.1.5).

Eq. (7.1) is the fundamental definition of the coefficient of friction as it was originally established by Amontons (see Section 3.2.2). As already discussed in Chapter 3, the quantity can be viewed as a normalised measure of the competition between shear stress promoting shear motion and normal stress impeding it. For simplicity, we will use the term 'friction' (either initial or static) to characterise the mechanical property of the surface; that is the specific value that the ratio shear stress/normal stress needs to overcome to trigger shear motion. Strictly speaking, the latter definition is physically incomplete but it is the most frequently adopted one amongst geologists. Slip tendency is independent of the mechanical properties of the surface; instead, it describes a particular aspect of the stress state applied to the surface, which indicates the tendency of the stress to promote shear motion along the surface. In summary, high friction lowers the probability for shear reactivation, whereas high slip tendency favours its potential occurrence.

The principles of slip tendency analysis can be easily explained with the help of Mohr diagrams. Considering usual geometrical relationships in the Mohr space, the slip tendency of any plane of the 3D space is conveniently related to the angle ω (Fig. 7.3A) such that

$$T_S = \tan\omega \tag{7.2}$$

T_{Smax} is determined for the maximum value of ω, and the pole of the corresponding plane is the point of tangency between the largest Mohr circle and a line passing through the origin. $T_{Smin} = 0$ is obtained for $\omega = 0$; in other words, the isoline $T_{Smin} = 0$ corresponds to the abscissa in the Mohr space (Fig. 7.3B) and contains the poles of the three planes parallel to the three principal planes of stress, i.e. the planes for which shear stress vanishes (see Section 3.1.4 for demonstration). Intermediate slip tendency isolines are built by simple rotation of a line, involving the origin, from $\omega = 0$ to $\omega = \tan^{-1} (T_{Smax})$. The procedure permits to attribute a unique slip tendency value to each planar discontinuity in the 3D space. Finally, the results are reported on stereograms for easier visualisation (Fig. 7.4).

The last refinement of the analysis consists of evaluating the results with respect to assumed or independently determined frictional properties of the studied discontinuities (Fig. 7.3B). This last step is similar to the traditional approach described in Section 3.2.2 (e.g. Fig. 3.14D) but presents the additional advantage of quantification. Planes whose poles plot above the friction line in the Mohr space are susceptible to slip and their susceptibility to slip is given by T_s.

Before pursuing our discussion, let us derive the analytical expression of slip tendency. Substituting Eq. (3.56) in Eq. (3.55), the formula for the normal stress becomes

$$\sigma_n = l^2\sigma_1 + m^2\sigma_2 + (1 - l^2 - m^2)\sigma_3 \tag{7.3}$$

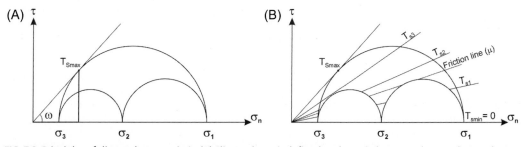

FIG. 7.3 Principles of slip tendency analysis. (A) Slip tendency is defined as the ratio between shear and normal stress or the tangent of the angle ω. Maximum slip tendency, T_{Smax}, is attached to the plane whose pole corresponds to the point of tangency between the Mohr circle and a line through the origin. (B) Slip tendency values for all other planes of the 3D space are determined by simple clockwise rotation of the tangent line, i.e. by decreasing ω, and eventually reaches $T_{Smin} = 0$ when the line becomes parallel to the abscissa. Shear motion is predicted for T_s values higher than the coefficient of friction (μ) of the planar discontinuities (note that the method is not restricted to cohesionless discontinuities).

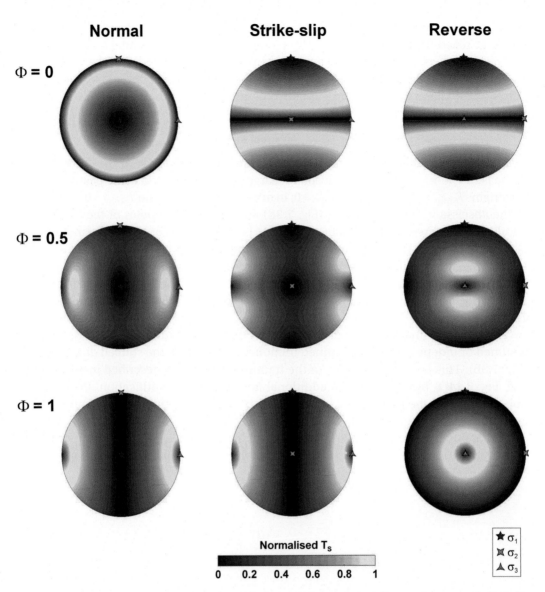

FIG. 7.4 Slip tendency (T_s) distributions for the three stress regimes as a function of stress anisotropy, as indicated by the shape ratio, Φ. Note that the distributions adopt girdle shapes for extreme values of Φ (i.e. uniaxial stress states). The plots were generated using the FracTend code of David Healy, University of Aberdeen, UK (Stephens et al., 2018) with $\sigma_3 = 10\,\text{MPa}$ and $\sigma_1 = 60\,\text{MPa}$. Slip tendency is normalised with respect to T_{Smax}. Projection is on the lower hemisphere of a Schmidt net.

where, by convention, l and m represent the direction cosines of the pole of the plane with respect to the maximum and the intermediate principal axes of stress, respectively (see Section 3.1.4 and Fig. 3.7).

Let us rewrite Eq. (7.3) in order to express σ_n as a function of Φ, the shape ratio given in Eq. (3.38):

$$\sigma_n = l^2(\sigma_1 - \sigma_3) + m^2(\sigma_2 - \sigma_3) + \sigma_3 \tag{7.4}$$

and finally

$$\sigma_n = (\sigma_1 - \sigma_3)(l^2 + m^2\Phi + Q) \tag{7.5}$$

where

$$Q = \frac{\sigma_3}{\sigma_1 - \sigma_3} \tag{7.6}$$

Now let us find a convenient equation for the shear stress. According to Eq. (3.70), the square of τ is

$$\tau^2 = (\sigma_1 - \sigma_3)^2 \left[l^2 \left(n^2 + m^2(1 - \Phi) \right)^2 + m^2 \left(n^2\Phi - l^2(1 - \Phi) \right)^2 + n^2 \left(m^2\Phi + l^2 \right)^2 \right] \tag{7.7}$$

Squaring Eq. (7.5) and combining it with Eqs (7.1) and (7.7), we get

$$T_S^2 = \frac{l^2 \left(n^2 + m^2(1 - \Phi) \right)^2 + m^2 \left(n^2\Phi - l^2(1 - \Phi) \right)^2 + n^2 \left(m^2\Phi + l^2 \right)^2}{\left(l^2 + m^2\Phi + Q \right)^2} \tag{7.8}$$

Re-organising Eq. (7.8), which is somewhat laborious but does not present any serious difficulty, we obtain

$$T_S^2 = \frac{m^2 \left(l^2 + n^2 \right)\Phi^2 - 2l^2 m^2\Phi + l^2 \left(m^2 + n^2 \right)}{\left(l^2 + m^2\Phi + Q \right)^2} \tag{7.9}$$

which, according to Eq. (3.43), can be rewritten in a more pleasant form:

$$T_S^2 = \frac{m^2 \left(1 - m^2 \right)\Phi^2 - 2l^2 m^2\Phi + l^2 \left(1 - l^2 \right)}{\left(l^2 + m^2\Phi + Q \right)^2} \tag{7.10}$$

Let us examine Eq. (7.10). Slip tendency depends on (1) the attitude of the plane (i.e. the three direction cosines), whose influence is evident in Fig. 7.3, (2) stress anisotropy (i.e. Φ) and (3) the quantity Q, which, according to Eq. (7.6), is the ratio of σ_3 on the differential stress. The latter quantity measures the relative contributions of pressure and differential stress on slip tendency (Fig. 7.5A and B). For an increase in σ_3 (differential stress remaining constant), the Mohr circle is shifted to the right and, therefore, the mean stress (or pressure) increases, resulting in higher normal stresses on all planes and, thus, lower slip tendencies. In contrast, increasing differential stress (Fig. 7.5B) implies higher shear stress on all planes (except for the planes perpendicular to the principal axes of stress) and, thus, higher slip tendencies. The way slip tendency varies in function of stress anisotropy is not directly apparent from Eq. (7.10) but becomes clear in the Mohr space (Fig. 7.5C). When Φ increases, slip tendency increases for planes orthogonal to the $(\vec{\sigma}_2, \vec{\sigma}_3)$ plane while it decreases for planes orthogonal to $(\vec{\sigma}_1, \vec{\sigma}_2)$. The latter behaviour is more evident in the stereograms depicted in Fig. 3.4.

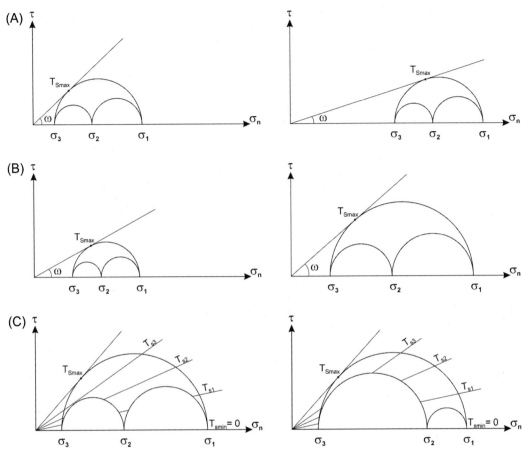

FIG. 7.5 Variation in average slip tendency as a function of the stress parameters involved in Eq. (7.10); see text for details. (A) Increase in σ_3 while maintaining constant differential stress (i.e. increase in mean stress) results in a decrease in average slip tendency, conveniently depicted here by the decrease in maximum slip tendency, T_{Smax}, and, in particular, the tightening of ω. (B) Increase in differential stress results in an increase in average slip tendency (i.e. opening of ω). (C) Increase in shape ratio, Φ (suggested here by the relative position of σ_2), promotes an increase in average slip tendency for planes perpendicular to the $(\vec{\sigma}_2, \vec{\sigma}_3)$ plane but a decrease for planes perpendicular to the $(\vec{\sigma}_1, \vec{\sigma}_2)$ plane (see also Fig. 7.4).

To complete our analysis, let us determine the values for which Eq. (7.10) reaches its extreme values, $T_{Smin} = 0$ and T_{Smax}. Fig. 7.5C shows that neither T_{Smin} nor T_{Smax} depend on Φ but solely on plane attitude (rigorous demonstration of the latter is straightforward but would lengthen the discussion unnecessarily). Therefore, Eq. (7.10) needs to be treated as a function of the director cosines and, consequently, accepts four roots. Two roots are evident from Eq. (7.10), and both correspond to $T_{Smin} = 0$: $l = 1$, that is, σ_1 perpendicular to the plane and thus $m = 0$, and $l = 0$, that is, σ_1 parallel to the plane and thus $m = 0$ or $m = 1$. All three cases describe the three situations previously identified where the plane parallels one of the three principal planes of stress.

The pole of the plane for which T_S reaches its maximum T_{Smax} belongs obligatorily to the $(\vec{\sigma}_1, \vec{\sigma}_3)$ plane (Fig. 7.3) or accepts $m = 0$ as director cosine. The latter particularity allows for a simplification of the equation to be treated; setting $m = 0$ in Eq. (7.10) leads to

$$T_S^2 = \frac{l^2(1-l^2)}{(l^2+Q)^2} \tag{7.11}$$

which is the equation giving T_S for the poles in the $(\vec{\sigma}_1, \vec{\sigma}_3)$ plane and, expectedly, is independent of the value taken by the shape ratio.

Assuming Q is known, T_{Smax} is determined after differentiating Eq. (7.11) with respect to l and searching for the zeros of the obtained function. Once again, the analytical development does not present any major difficulty; nevertheless, it is advised to differentiate the logarithm of the equation instead of differentiating brutally the equation itself...

The development leads to the expression of the two roots (i.e. direction cosines) corresponding to T_{Smax}:

$$l_{max}^2 = \frac{Q}{2Q+1} \tag{7.12}$$

Similar to what has been discussed in Section 3.1.5 (e.g. Fig. 3.9B), the solutions describe two symmetrical planes but with opposite shear stress values; thus, neglecting the negative root does not affect the main outcome of the analysis.

Note that in the present case, l is also the cosine of the angle θ introduced to define 2D Mohr diagrams in Section 3.1.5. Thus, Eq. (7.12) can be rewritten as

$$\theta_{max} = \cos^{-1}\left(\sqrt{\frac{Q}{2Q+1}}\right) \tag{7.13}$$

As the final step, T_{Smax} is calculated by combining Eqs (7.11) and (7.13).

b) Extension of fault slip tendency analysis to the inverse problem

Fault slip tendency analysis is conducted as a forward problem traditionally. Only the two recent works by McFarland et al. (2012) and Morris et al. (2016) proposed to extend the theory to the inverse problem of paleostress determination. We showed in the previous section that spatial distribution of slip tendency depends on orientations of the principal axes of stress, stress anisotropy and stress magnitude (Fig. 7.4 and Eq. 7.10). The core problem to be solved consists of finding an observable that relates directly to slip tendency and permits to map its spatial distribution.

McFarland et al. (2012) advanced fault displacement as a proxy for fault slip tendency (Fig. 7.6). Intuitively, faults with high slip tendency during the time of application of a given stress state are expected to be often reactivated and thus should accumulate relatively large displacements. In contrast, faults with low slip tendency are anticipated to show relatively small displacements. However, faults with high slip tendency could have

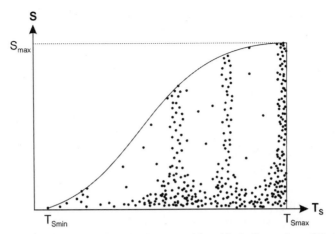

FIG. 7.6 The assumed relationship between fault displacement (S) and fault slip tendency (T_S) after McFarland et al. (2012) and Morris et al. (2016). The hypothetical envelope represents maximum cumulated displacements during the tectonic or stress event as a function of T_S. Faults with high T_S may or may not accumulate large displacements, whereas faults with low T_S accumulate modest displacements only (see text for explanation). Black dots represent synthetic data points. Note that T_{Smin} corresponds to $S > 0$. The envelope has no precisely defined mathematical shape in general and is depicted schematically.

formed at any time of the stress event and potentially bear modest displacements if initiated tardily. In summary, fault orientations associated with both significant and modest displacements point to high T_S values, whereas fault orientations with exclusively low amounts of displacement indicate low T_S values (Fig. 7.6).

However, translating the latter qualitative relationships in mathematical terms constitutes the most difficult part in the elaboration of the method. Solving the problem requires normalisation of the two quantities to be compared during the computations, i.e. fault displacement, S, and slip tendency, T_S. After numerous attempts, McFarland et al. (2012) selected the following two normalised quantities:

$$\overline{S} = \frac{S}{S_{max}} \tag{7.14}$$

where S_{max} is the maximum fault displacement of the dataset (i.e. measured fault attitudes and related displacements), and

$$\overline{T}_S = \frac{T_S - T_{smin}}{T_{smax} - T_{smin}} \tag{7.15}$$

where T_{Smax} and T_{Smin} are the maximum and minimum slip tendencies, respectively, as computed for the dataset.

The normalised values are utilised to define an empirical discontinuous penalty function, ε, such that

$$\varepsilon = -0.5\left(\overline{S} - \overline{T}_S\right)^2 \text{ for } \overline{S} \leq \overline{T}_S \tag{7.16}$$

$$\varepsilon = 2\left(\overline{S} - \overline{T}_S\right) \text{ for } \overline{S} > \overline{T}_S \tag{7.17}$$

The penalty function accounts for the hypothetic distribution depicted in Fig. 7.6. No penalty is imposed on the optimal case, $\overline{S} = \overline{T}_S$. The penalty is moderate for displacements lower than predicted (Eq. (7.16)), which according to the previous discussion remain realistic, but severe if higher than expected (Eq. 7.17).

However, the results of the method are strongly dependent on fault orientation. For example, a tight cluster of fault poles poorly constrains the whole slip tendency pattern on a stereogram, as a comparison of the different patterns depicted in Fig. 7.4 suggests. To mitigate the impact of this potential drawback, McFarland et al. (2012) proposed to add 'zero-displacement artificial data' to the actual data. The procedure consists of adding, either manually or by means of mathematical extrapolation, zero-displacement fault planes, which in practice are difficult to identify and are seldom included in natural datasets. In turn, the artificial data permit to fill critical gaps in the spatial distribution of fault slip tendency. A last refinement of the penalty function is made to treat these artificial data:

$$\varepsilon = 2T_S^2 \text{ for } \overline{S} = 0 \tag{7.18}$$

Once the penalty function is established, the best-fitting paleostress tensor is determined after minimising the function:

$$F = \sum_{i=1}^{N} \varepsilon_i^2 \tag{7.19}$$

where i is a dummy index and N is the total number of fault displacement data, including the 'zero-displacement artificial data'.

The minimisation is performed with the help of traditional strategies (see McFarland et al., 2012 for details), and uncertainties are estimated by means of Bayesian analysis.

The method of McFarland et al. (2012) opens new and interesting perspectives to paleostress analysis. However, at the date the present lines are written, it is difficult to evaluate in detail the strengths and weaknesses of this recent method. The two published works of McFarland et al. (2012) and Morris et al. (2016) furnish only some elements for discussion. In contrast to traditional fault slip inversion methods (Chapter 4), the approach relies on (amount of) fault displacement data. It thus potentially extends paleostress analysis to faults void of kinematic markers or imaged in the subsurface (see Section 7.2 for further discussion on the topic). However, it is generally extremely difficult to estimate the total fault displacement without constraining the associated slip vector, suggesting that the method is safely applicable to particular cases only.

Comparison between the results delivered by a classical method (i.e. the right trihedra method of Lisle, 1987; see Section 4.3.3) and the method of McFarland et al. (2012) showed good agreement in terms of orientations for the principal axes of stress (Morris et al., 2016). The latter outcome seemed to validate the background assumption relating displacement to slip tendency. However, the benchmarking exercise returned large discrepancies between the shape ratio values delivered by both methods, i.e. $\Phi = 0.32$ and $\Phi = 0.28$ for McFarland's method, and $\Phi = 0.90$ and $\Phi = 0.85$ for Lisle's method.

Morris et al. (2016) did not describe the fault slip data they analysed. However, qualitative examination of the stereograms presenting the poles of the inverted fault planes (fig. 8 in Morris et al., 2016) reveals both sufficiently high amounts of data and satisfactory variety in fault orientations, suggesting that the results of the paleostress determination with the right trihedra method were robust. These considerations suggest that the method of McFarland et al. (2012) fails to resolve accurately stress anisotropy. One might speculate that the addition of 'zero-displacement artificial data' does not solve completely the problem of data gaps and even introduces a systematic bias, but further testing of the method is required before drawing firm conclusions. The approach of McFarland et al. (2012) is much too recent and, based on the scarce knowledge at hand, assessing its degree of validity is a rather risky task, but admittedly its originality is appealing.

7.2 Paleostress reconstructions from fault slip sense data

7.2.1 Dip separation and dip-slip

Geologists commonly face situations where the faults can be detected owing to clearly expressed offsets but where neither kinematic markers nor friction striae are visible, impeding fault slip determination and, a priori, the use of traditional fault slip inversion methods. Such a situation is met when interpreting subsurface faults as imaged by geophysical methods (e.g. Fig. 7.1) but is also relatively frequent in the field when fault mirrors lack slickenside kinematic indicators.

One may measure, alternatively, *fault separation* and, more specifically, *dip separation* (Fig. 7.7), that is the separation along the dip line of the fault of the two linear traces of a

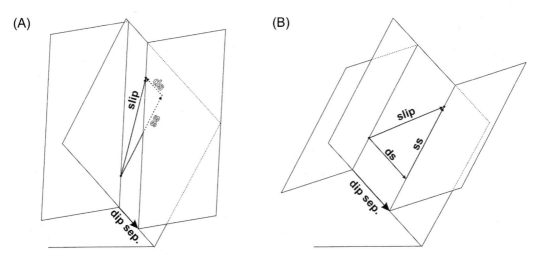

FIG. 7.7 (A) In general, the sense of dip separation (i.e. dip sep.) does not constrain at all fault slip; the case depicted here shows that dip separation can even contradict the actual sense of the dip-slip component. (B) Dip separation sense systematically agrees with dip-slip sense if the cut-off lines are horizontal (i.e. the marker plane and the fault share the same strike or the marker plane is horizontal). ds and ss stand for dip-slip and strike-slip, respectively.

marker plane (i.e. *cut-off lines* or *cut-off*). Dip separation furnishes an apparent sense of fault motion, but it is not straightforward to extract relevant information on fault slip from the sense of dip separation. Dip separation remains an ambiguous element of observation to constrain fault slip when the fault and the marker plane differ in strike orientation. In the worst case, the sense of dip separation (i.e. normal in Fig. 7.7A) can even oppose the sense of the dip-slip component (i.e. reverse in Fig. 7.7A). However, both senses agree systematically when the cut-off lines are horizontal (Fig. 7.7B) or, in other words, when both the fault and the marker plane strike parallel or when the marker plane is horizontal—the latter configuration being common in sedimentary basins.

The original inversion method of Lisle et al. (2001) exploits the systematic relationship between dip separation and dip-slip, described above, in structural settings conforming with its validity. By definition, the inversion relies only on fault plane attitudes and their corresponding dip-slip components. Hence, the characteristic information on the nature of fault slip as gathered in each datum is reduced to two attributes: 'normal' and 'reverse'. Therefore, the goal of the fault slip inversion is to find the reduced stress tensor that best explains the spatial distribution of normal and reverse dip-slips. The latter idea stems from the astute approach of Davidson and Park (1978), who inferred graphically limiting orientations for σ_1 and σ_3 from sheared dykes measured in Western Greenland and showing either normal or reverse dip-slip.

7.2.2 Slip sense criterion and properties of the stress quadric of Cauchy

It is technically possible to use one of the classical fault slip inversion methods (i.e. the ones described in Chapter 4) and compute the reduced paleostress tensor that best explains the dip-slip data. However, owing to the incomplete nature of the input, the unique result that this type of method usually delivers represents an approximation whose robustness is difficult to evaluate. To tackle this problem, the numerical inversion of 'imperfect' dip-slip data demands efficient routines capable to compute series of paleostress tensors in reasonable times and detection of the most likely solution from a relatively large collection of equally acceptable tensors.

The fast algorithm developed by Lisle et al. (2001) and Orife et al. (2002) is based on fundamental properties of the *stress quadric of Cauchy* (e.g. Ramsay and Lisle, 2000, pages 791–792). A second-rank tensor, like the stress tensor, can be represented by a second-order surface, i.e. a *quadric*, whose general equation is

$$\lambda_1 x_1^2 + \lambda_2 x_2^2 + \lambda_3 x_3^2 = 1 \tag{7.20}$$

where λ_1, λ_2 and λ_3 are the eigenvalues of the tensor and x_1, x_2 and x_3 are the Cartesian coordinates of a point of the quadric surface.

The quadric may adopt different shapes (e.g. hyperboloid or ellipsoid) depending on the signs taken by the eigenvalues. In the relatively common case of symmetric stress tensors and if all the three eigenvalues are positive, the quadric is an ellipsoid. Therefore, provided that

$$\sigma_1 \geq \sigma_2 \geq \sigma_3 > 0 \tag{7.21}$$

which represents the stress condition for 'pure' faulting in the brittle crust (i.e. the walls of the fracture remain in contact during slip), the stress quadric describes an ellipsoid (Fig. 7.8A), whose equation is

$$\sigma_1 x_1^2 + \sigma_2 x_2^2 + \sigma_3 x_3^2 = 1 \tag{7.22}$$

One should pay much attention to the fact that the stress quadric is not the stress ellipsoid presented in Chapter 3 (e.g. Fig. 3.6), though both objects are geometrically similar! Setting alternatively to zero two of the three coordinate values in Eq. (7.22) demonstrates that the respective lengths of the three axes of the quadric are equal to $\frac{1}{\sqrt{\sigma_1}}$, $\frac{1}{\sqrt{\sigma_2}}$ and $\frac{1}{\sqrt{\sigma_3}}$, the shortest and the longest axis being parallel to σ_1 and σ_3, respectively (Fig. 7.8A). Any radius, R, of the ellipsoid is therefore equal to (see e.g. p. 68 in Durelli et al., 1958 for complete mathematical treatment)

$$R = \frac{1}{\sqrt{\sigma_n}} \tag{7.23}$$

A powerful feature of the quadric representation of a second-order tensor resides in its radius-normal property. Concerning the stress quadric of Cauchy, the property can be summarised as follows: each point of the surface represents a plane of the physical space, whose normal is parallel to the radius R, connecting the point to the centre of the ellipsoid, and whose associated stress vector is perpendicular to the surface of the ellipsoid, at the point under consideration (Fig. 7.8A). In other words, the stress quadric representation allows determining graphically the orientation of the stress vector for any plane of the

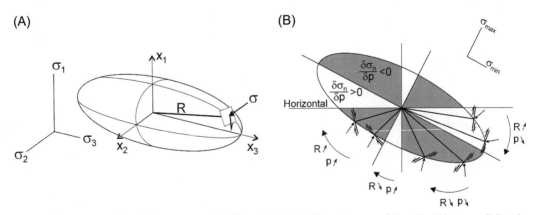

FIG. 7.8 (A) Representation of the stress quadric of Cauchy. By definition, the axes of the ellipsoid are parallel to the principal axes of stress; each point of the surface represents a plane of the physical space, the radius, R, and the stress vector, σ, being orthogonal to the plane and the ellipsoidal surface, respectively. Thus, the sense of slip can be determined from the geometrical representation (see text for details). Note that the longest/shortest axis is parallel to the axis of minimum/maximum principal stress. (B) Vertical section across the stress quadric depicting the relationship between slip sense and the relative variation of R (alternatively the relative variation of the normal stress, σ_n; see text for clarification) with respect to that of the dip, p. The arrows perpendicular to the trace of the quadric represent projected stress vectors. Grey areas indicate normal slip and correspond to negative values of the partial derivative of σ_n with respect to p, whereas white areas indicate reverse slip and correspond to positive values.

physical space; a goal impossible to reach based on the stress ellipsoid solely, save for the particular cases the plane corresponds to one of the three principal planes of stress (see Section 3.1.3). Finally, knowing the orientation of both the normal and the stress vector, the quadric representation furnishes automatically the shear stress vector and thus the slip sense on the plane.

Another agreeable property, a corollary of the previous one and instrumental in the method of Lisle et al. (2001), is the relationship between slip sense and variation of radius, R, with respect to variation of plane's dip, p. Let us imagine a vertical cross-section through the centre of the quadric in order to examine without too much effort the latter property, as evidenced on the elliptical trace (Fig. 7.8B). The very first remark is that this representation does not depict either the stress vectors or the principal axes of stress (save the vertical one) but their respective projections on the section plane. Nevertheless, slip sense, which is the key information needed by the method, remains untouched. R varies, i.e. increases or decreases, when it is rotated away from one axis of the ellipse to the other (that once again do not correspond to the axes of the quadric in the general case), whereas p varies with respect to the orientation of the horizontal; thus, the respective signs of relative variation of both quantities are not obligatory correlated. However, as shown in Fig. 7.8B, the sign of the derivative of R with respect to p correlates with slip sense such that

$$\text{for normal slips } \frac{\partial R}{\partial p} > 0 \tag{7.24}$$

$$\text{and for reverse slips } \frac{\partial R}{\partial p} < 0 \tag{7.25}$$

According to Eq. (7.23), these two conditions can be respectively re-formulated as

$$\text{for normal slips } \frac{\partial \sigma_n}{\partial p} < 0 \tag{7.26}$$

$$\text{and for reverse slips } \frac{\partial \sigma_n}{\partial p} > 0 \tag{7.27}$$

In brief, the sign of the derivative of the normal stress with respect to the plane's dip is a slip sense criterion and the criterion can be used to evaluate the goodness of fit of a stress tensor to a collection of fault slip sense data.

7.2.3 Numerical implementation and robustness of the method

The numerical method of Lisle et al. (2001) is a grid-search method (see Section 4.2.2), where a large number of tensors, typically thousands, are systematically tested. Despite the large number of trial tensors required to achieve reliable results, the simplicity of the background mathematics ensures short calculation times. In practical terms, the algorithm proceeds according to the following steps: (1) one trial tensor is selected; (2) normal stresses are calculated for all input fault planes using Eq. (3.55); (3) their derivatives with respect to dip are computed; (4) the slip senses are determined as a function of the signs of the derivatives associated to their respective fault planes (i.e. Eqs 7.26 and 7.27); (5) the simulated slip senses are compared to the input ones; (6) if a certain percentage (as fixed by the user, e.g. 90%) of simulated slip senses agree with the data, the stress tensor is a potential solution and is stored; (7) the stress parameters are modified and the previous steps are reiterated. Evidently, step (3) demands the introduction of an additional sub-step, where 'shadow' fault planes are generated by slightly incrementing the dip of each fault (i.e. by ~0.05°; Orife et al., 2002), and normal stresses and, ultimately, derivatives are computed. The computation stops when a sufficient number of trial tensors cover the whole parameter space with acceptable accuracy.

As already pointed out, due to the inherent loss of information content of the data, the results of the fault slip sense inversion method are expected to be less accurate compared to those obtained using classical paleostress inversion programs. Statistical treatment of the results (i.e. collections of best-fitting reduced stress tensors) may mitigate the latter drawback. Fig. 7.9 depicts two benchmarking tests of the method. In the first example (Fig. 7.9A), Lisle et al. (2001) simulated artificial fault slip data by directly computing slip vectors for a set of planes with variable orientations and according to selected stress parameters (i.e. similarly to the approach discussed in Section 7.1.1). After having removed strike-slip components, the data were inverted. The analysis revealed that the average orientations of the principal axes of stress of the 13 best-fitting tensors were in good-to-excellent agreement with the axis of the input tensor. It also showed that the modal value of the 13 computed shape ratios was very close to that of the initial reduced tensor.

The second example (Fig. 7.9B) compares the result of a traditional paleostress reconstruction, based on fault slip data measured in the field, with the ones obtained after inversion of the same dataset downgraded to its dip-slip components (Orife et al., 2002). The fault slip sense data inversion delivered 4645 acceptable tensors. Statistically, the orientations of the principal axes of stress and the shape ratios showed good agreement with the values obtained from inversion of the full fault slip components.

Benchmarking gives a hint on the goodness of the method but fails to furnish any formalised quantitative measure of its precision, i.e. on the spread of the results. To address the latter issue, Lisle et al. (2001) and Orife et al. (2002) introduced the 'solution percentage', which they defined as the ratio between the number of solution tensors and the total number of tested tensors. Ideally, the smaller the solution percentage, the more precise the determination.

For instance, a small number of fault slip data does not constrain the solution well; therefore, the data can be equally explained by a relatively high number of stress tensors. Consequently, the solution percentage is high but would decrease if more data were involved in the computations (only if the added data do not replicate the information already present in the initial sample). The same line of reasoning applies to data variety. As we already pointed out in Chapter 4, robust paleostress determination requires fault slip data with orientation variety. The solution percentage is logically higher for a relatively uniform dataset, which automatically leads to a large number of solution tensors, than for

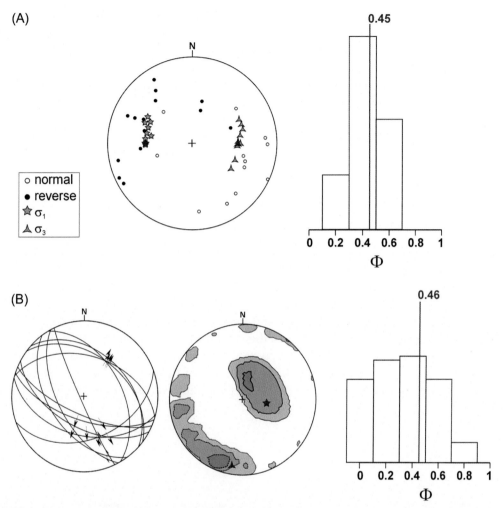

FIG. 7.9 Tests of the method of inversion of fault slip sense data. (A) Synthetic fault slip data (fault plane poles depicted as *open or solid circles*) were generated by means of forward computations with σ_1 and σ_3 oriented, as indicated by the *red star* and the *red triangle*, respectively, and $\Phi = 0.45$. As the second step, only the dip-slip components of the artificial data were kept and inverted to find the collection of best-fitting tensors. The results corresponding to the 13 best-fitting tensors are presented in the form of clouds of σ_1 and σ_3 poles (*grey symbols*) and frequency distribution of Φ values (after Lisle et al., 2001). (B) Fault slip data measured by Bellier et al. (1989) and results of their fault slip inversion are indicated by *red symbols and numbers*. *Grey contours* represent the results of the inversion of the dip-slip components of the same dataset, carried out by Orife et al. (2002), and correspond to σ_1 and σ_3 density contours (i.e. 3%, 5% and 7% per 1% area, and 2%, 3% and 4% per 1% area, respectively). The frequency distribution of Φ values for the 4645 acceptable tensors is also shown. Stereograms are lower hemisphere Schmidt nets.

a dataset containing a wealth of orientations, which favours tightening of the collection of acceptable tensors. However, too much variety in the dataset or too much data may conduct to the case where no solution at all is found, warning the user that the data may be heterogeneous.

The tests (Fig. 7.9) suggest that inversion of fault dip-slip senses provides reasonable estimates of paleostress parameters, although it involves incomplete characterisation of slips. Thus, it is the best approach at hand when dealing with subsurface data. To note, the validity of this type of methodology was demonstrated by Sato (2006) in his theoretical study.

7.3 Geomechanical approaches

7.3.1 Faulting and stress perturbations

Paleostress inversion by means of geomechanical numerical modelling is a brand new approach developed by Maerten (2010) and Kaven et al. (2011). Geomechanical approaches take advantage of stress perturbations related to faulting, whereas the inversion methods described in previous chapters neglect them.

In the first half of the 20th century and using the classical solution of Inglis (1913), Anderson calculated stress deflections in the vicinity of a highly elongated elliptical cavity submitted to shear (e.g. Anderson, 1951, p. 163). The calculations of Anderson (and of his numerous successors) predict progressive rotations of the principal axes of stress such that σ_1 tends to become either parallel or perpendicular to the discontinuity at its tips

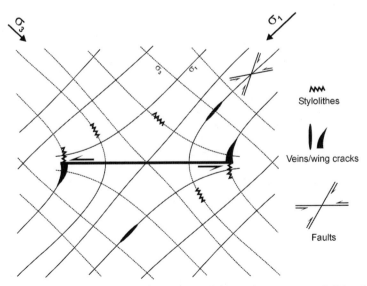

FIG. 7.10 Stress deflections related to faulting and some potential secondary structures. Solid and dashed lines indicate σ_1 and σ_3 trajectories, respectively (after Anderson, 1951).

(Fig. 7.10). In turn, when rock strength is exceeded, the geometries of the subsequent secondary structures (e.g. veins, stylolithes and/or minor faults) reflect the local stress directions but not the orientation of the 'remote' stress responsible for faulting along the main fault(s). This inference has been validated both from field observations (e.g. Segall and Pollard, 1980; Rispoli, 1981) and experimentally (e.g. Petit and Barquins, 1988) in the course of numerous studies. In short, if secondary structures can be unambiguously identified as coeval to the displacement along a given fault, these can be used to map the pattern of the stress trajectories associated with the faulting, which in turn indicates the characteristics of the remote stress. The latter statement is the essence of paleostress reconstructions based on geomechanical numerical modelling.

Numerical techniques capable of simulating the stress pattern related to a discontinuity submitted to shear displacement have existed for decades (e.g. the Distinct Element Method of Cundall, 1971). It was possible to run series of forward models, changing the orientation of boundary stress with respect to fault strike, until reaching satisfactory fit between the simulated stress trajectories and the orientations of the structures measured in the field (Homberg et al., 1997). However, these techniques were laborious and the results (i.e. knowledge of best-fitting remote stress) were rather frustrating as the exploration of the space of solutions was limited by the relatively low number of reasonably feasible trial-and-error modelling runs, especially in 3D. Moreover, the incorporation of a proper inversion scheme in mechanical modelling codes was not possible due to the limited performance of the computers at the time.

The emergence of a solution to the computing problem had to await the exponential improvement in computer speed and storage capacity that took place during the past two decades. Although the solving of the complete set of 3D mechanical equations, even for relatively simple linearly elastic models, remains challenging for desktop computers when numerous stress tensors are tested, Maerten (2010), Kaven et al. (2011) and their co-workers (Maerten et al., 2014, 2016a, 2016) succeeded to tackle the problem with a numerical method based on the boundary element method (BEM), namely iBem3D.

7.3.2 Numerical treatment of the inversion problem (iBem3D)

a) Principles of the inversion and boundary element method

The goal of the inversion is to determine the remote stress, whose resultant stress trajectories within the modelled volume honour the (apparent) displacements along the studied major faults and the attitudes (and potentially sense of shear) of coeval secondary structures (Fig. 7.11). To note, the method advanced by Maerten et al. is also capable of exploiting other datasets as input (e.g. GPS and InSAR). As mentioned previously, the treatment of the inversion problem requires some strategy to reduce computing time. The adopted strategy involves the use of the BEM. Additional features include reduction of the remote stress tensor and implementation of the principle of superposition.

Unlike classical numerical methods, like the finite element method (FEM), for which the solution is computed for each element of the modelled volume, the BEM solves the

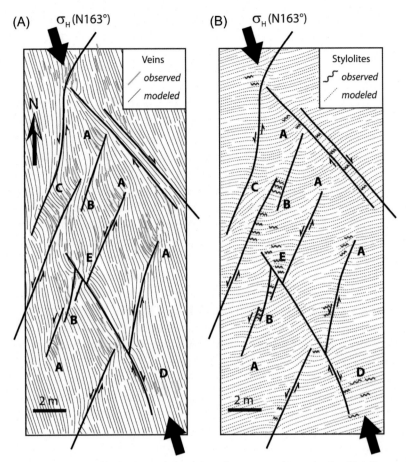

FIG. 7.11 Results of the geomechanical paleostress inversion of an outcrop located at Les Matelles, southern France, exhibiting coeval strike-slip faults, veins and stylolithes (adapted from Maerten et al., 2016). (A) Comparison between measured vein orientations and modelled maximum horizontal stress trajectories. (B) Comparison between orientations of measured stylolithic planes and modelled minimum horizontal stress trajectories. The inversion points to a strike-slip stress regime with σ_1 trending N163°E and $\Phi = 0.45$. *Letters* represent specific model locations discussed in Maerten et al. (2016). *Reprinted (in part) with permission from Elsevier.*

problem only at the boundaries of the modelled volume and the obtained solution, in the form of integral equations, is applied to compute the solutions for the rest of the volume in the post-processing phase (see e.g. Gaul et al., 2003 for further details). For example, if a traditional FEM were employed, the computing of stress trajectories in an elastic plate containing a hole and submitted to boundary stresses at its edges would require the meshing of the whole plate and the computation of the solution at each element of the mesh. In addition, the edges of the plate would need to be located far enough from the hole to avoid spoiling of the solution by edge effects. Therefore, a significant amount of computing time would be devoted to solve the mechanical equations at elements relatively distant from the hole and thus would be spent to deliver local solutions of little interest. When applying the BEM, the periphery of the hole is treated as a boundary and the equations are only solved for that specific boundary; once the boundary problem is solved, the solution is used to compute the stress field in the zones of interest for the user, without the need of dimensioning and meshing the plate on beforehand. The BEM presents a clear advantage in terms of computing time and is particularly well-adapted to handle problems involving internal boundaries, such as contacts in shear or, in geological terms, faults.

In iBem3D, the mechanical behaviour of faults is modelled based on the analytical solution for angular dislocations[a] in homogeneous or inhomogeneous linear elastic whole- or half-space (Maerten et al., 2014 and references therein). Fault walls are constructed using triangular elements, which are particularly convenient to design relatively complex fault surfaces. The construction allows for simulating discontinuous slip (i.e. variations in slip orientation according to local fault attitude and variations in slip magnitude according to distance from fault tips) but displacement normal to fault elements is prescribed to zero in order to avoid potential gap openings and interpenetration of the fault blocks. Both stress and displacements can be set as boundary conditions, though only stresses are relevant when the code is used for paleostress inversion. Following the principle of the BEM, displacement along the modelled fault is computed and then the solution is used to calculate displacement and strain at selected observation points within the fault blocks. Stress is finally derived from strain according to the predefined elastic parameters. Stress at each observation point is computed as the sum of the stress disturbed by the faulting and the boundary stress.

b) Reduction of the stress tensor

Although the code involves an iterative solver and parallel computing, the inversion demands computation of numerous stress configurations. Similar to traditional inversion methods, geomechanical inversion considers a reduced form of the stress tensor.

An Andersonian boundary stress condition is assumed as the very first simplification (Fig. 7.12A):

$$T_A = \begin{pmatrix} \sigma_h & 0 & 0 \\ 0 & \sigma_H & 0 \\ 0 & 0 & \sigma_V \end{pmatrix} \tag{7.28}$$

By convention, σ_h and σ_H, respectively, indicate the minimum and maximum principal axes of stress in the horizontal plane, and σ_V the principal axis of stress according to vertical.

As for the procedure discussed in Section 4.1.3, the reduction of the stress tensor is conducted in splitting it into two components:

---

[a]Dislocation theory was originally introduced by the Italian mathematician Vito Volterra (1860–1940) in 1907 (Volterra, 1907) and mostly applied to plastic deformation of crystals before being used to model faulting (Steketee, 1958a, 1958b).

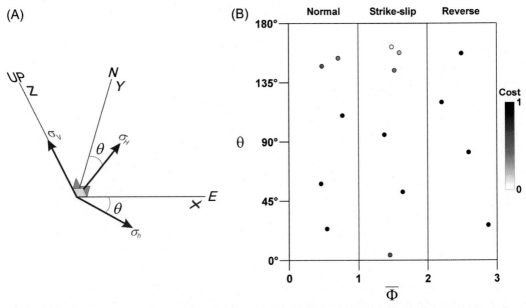

FIG. 7.12 (A) Andersonian stress state selected for the geomechanical numerical inversion and notation conventions used in the text; σ_h and σ_H are, respectively, the minimum and maximum principal axes of stress in the horizontal plane, and σ_V is the vertical principal axis of stress. (B) Fictitious example of results obtained after inversion as represented on a 2D diagram, where θ is the angle defined in (A) and $\overline{\Phi}$ the particular form of the shape ratio introduced by Lejri et al. (2015). *The circles* represent trial Andersonian stress states, and *light and dark tones* represent their associated computed costs: the lower the cost, the better the fitness of the tensor. The optimal stress tensor (*white circle*) corresponds to a strike-slip regime with $\theta \sim$N160°E and $\overline{\Phi} \sim$1.5 (i.e. $\Phi \sim 0.5$). See text for detailed explanations.

$$T_A = \begin{pmatrix} \sigma_h - \sigma_V & 0 & 0 \\ 0 & \sigma_H - \sigma_V & 0 \\ 0 & 0 & 0 \end{pmatrix} + \sigma_V I \tag{7.29}$$

where I is the unit matrix.

Retaining the component relevant to predict shear stress orientations, we get

$$T_{AD} = \begin{pmatrix} \sigma_h - \sigma_V & 0 & 0 \\ 0 & \sigma_H - \sigma_V & 0 \\ 0 & 0 & 0 \end{pmatrix} \tag{7.30}$$

Before proceeding further with the reduction, let us recall the expression of the stress tensor in the coordinate system of the physical space (see also Chapter 4, Section 4.2.2.a), the one of interest for the computations:

$$T_{xyz} = \Omega T_{AD} \Omega^T \tag{7.31}$$

where Ω is the rotation matrix and the superscript T indicates matrix transpose.

The choice of an Andersonian stress state simplifies the form of Ω, which can be written as

$$\Omega = \Omega_\theta = \begin{pmatrix} \cos\theta & \sin\theta & 0 \\ -\sin\theta & \cos\theta & 0 \\ 0 & 0 & 1 \end{pmatrix} \qquad (7.32)$$

where θ represents σ_H trend (Fig. 7.12A).

Considering the specific form of T_{AD} (Eq. 7.30), where the third row and the third column exclusively contain zeros, the system of equations can be simplified by subtracting one dimension to the parameter space, and thus the stress tensor and the rotation matrix can be rewritten as

$$T'_{AD} = \begin{pmatrix} \sigma_h - \sigma_V & 0 \\ 0 & \sigma_H - \sigma_V \end{pmatrix} \qquad (7.33)$$

$$\text{and } \Omega'_\theta = \begin{pmatrix} \cos\theta & \sin\theta \\ -\sin\theta & \cos\theta \end{pmatrix} \qquad (7.34)$$

The last step of the reduction of the stress tensor consists of substituting σ_1, σ_2 and σ_3 in Eq. (7.33) while paying attention to the stress regime and dividing T'_{AD} by the differential stress, $\sigma_1 - \sigma_3$. The latter operations result in three expressions of the reduced stress tensor, all three depending on the shape ratio Φ:

$$T'_R = \begin{pmatrix} -1 & 0 \\ 0 & \Phi - 1 \end{pmatrix} \text{ for normal stress regime} \qquad (7.35)$$

$$T'_R = \begin{pmatrix} -\Phi & 0 \\ 0 & 1 - \Phi \end{pmatrix} \text{ for strike} - \text{slip stress regime} \qquad (7.36)$$

$$\text{and } T'_R = \begin{pmatrix} \Phi & 0 \\ 0 & 1 \end{pmatrix} \text{ for reverse stress regime} \qquad (7.37)$$

For the sake of convenience, the traditional shape ratio Φ is rewritten as follows (e.g. Lejri et al., 2015):

$$\text{For normal stress regime } \overline{\Phi} = \Phi \quad \overline{\Phi} \in [0; 1] \qquad (7.38)$$

$$\text{For strike} - \text{slip stress regime } \overline{\Phi} = 2 - \Phi \quad \overline{\Phi} \in [1; 2] \qquad (7.39)$$

$$\text{For reverse stress regime } \overline{\Phi} = 2 + \Phi \quad \overline{\Phi} \in [2; 3] \qquad (7.40)$$

$\overline{\Phi}$ varies continuously from 0 to 3, integer values representing uniaxial stress states, and together with θ, the two values allow for characterising all possible Andersonian stress states and for easy reading of the results of the inversion on a 2D diagram (Fig. 7.12B).

Finally, the reduced stress tensor expressed in the coordinate system of the physical space is

$$T'_{xyz}(\theta, \overline{\Phi}) = \Omega'_\theta T'_R \Omega'_\theta T \qquad (7.41)$$

In summary, the Andersonian assumption adopted by Maerten et al. (2016) leads to a drastic reduction in the number of unknowns and subsequent reduction in computation time and memory use. Instead of dealing with the six unknowns of the stress tensor (which would be technically feasible), the computations involve only two unknowns: the trend of the maximum principal stress axis in the horizontal plane and the shape ratio.

c) Implementation of the principle of superposition

In addition to the choice of the BEM and of the reduction of the stress tensor, the high number of simulations required for solving the problem accurately (i.e. typically thousands to tens of thousands of simulations) are treated by further optimisation of the algorithm. The physical system under scope is linear, which means that if multiple independent loads are simultaneously applied to it, each individual load does not alter the effects of the other loads. Alternatively, one can consider an applied load as the sum of independent loads, whose effects can be treated individually. Therefore, both the boundary conditions and the internal solutions can be broken down into sums of independent boundary conditions and solutions respectively. The latter constitutes the *principle of superposition* extensively employed to solve a large variety of physical problems (Brillouin, 1946).

In the present case, any reduced stress tensor applied as boundary condition can be separated into three independent components:

$$T'_R = \alpha_1(\theta, \overline{\Phi}) T'_{R1} + \alpha_2(\theta, \overline{\Phi}) T'_{R2} + \alpha_3(\theta, \overline{\Phi}) T'_{R3} \tag{7.42}$$

where T'_{Ri} with $i \in [1;3]$ represents three independent Andersonian stress states and $\alpha_i(\theta, \overline{\Phi})$ are real numbers depending exclusively on the two stress parameters θ and $\overline{\Phi}$.

In virtue of the principle of superposition, any 'effect' inside the considered volume deriving from the application of T'_R can be separated into three corresponding components:

$$f = \alpha_1(\theta, \overline{\Phi}) f_1 + \alpha_2(\theta, \overline{\Phi}) f_2 + \alpha_3(\theta, \overline{\Phi}) f_3 \tag{7.43}$$

where f represents displacement, strain or stress.

According to Eq. (7.42), any Andersonian stress state can be computed (and thereafter expressed in the physical coordinate system) from the knowledge of three arbitrary prescribed but independent stress states. Thus, all the parameter space can be explored by varying the three $\alpha_i(\theta, \overline{\Phi})$ coefficients once these three initial stress states are selected. According to Eq. (7.43), displacement, strain and stress at any point of the model can be computed from the knowledge of the solutions obtained after having successively applied T'_{R1}, T'_{R2} and T'_{R3} (i.e. f_1, f_2 and f_3), and from the values chosen for three $\alpha_i(\theta, \overline{\Phi})$ coefficients.

Practically, the standard BEM computations are performed for only three initial stress states, avoiding re-computing of the whole model for each trial stress state of the parameter space. The solutions corresponding to the remaining (tens of) thousands of stress tensors to be tested are directly derived from Eq. (7.43). The use of the principle of superposition allows for considerable reduction of the total number of operations and consequently of computation time as only 18 multiplications are involved in the calculation of e.g. the local stress tensor at a selected point of the model.

To note, 3D displacement, strain and stress are computed, though the input reduced stress tensor is defined in two dimensions (Eqs 7.35–7.37). For example, local stress perturbations might produce the departure of one of the principal axes of stress from vertical; thus, the six components of the local stress tensor must be computed at each boundary element and at each point of interest inside the model.

The final step of the computations is to discriminate the best-fitting stress tensor amongst the collection of the tested tensors, i.e. to determine the two optimal stress parameters θ and $\overline{\Phi}$ or, in agreement with Eq. (7.43), the three corresponding $\alpha_i(\theta, \overline{\Phi})$ coefficients.

d) Fitness of the results

The best-fitting stress tensor is the one that promotes a stress field consistent with most fracture and stylolithe data or, ideally, consistent with all of them (Fig. 7.10). Thus, the quantitative criteria for estimating the goodness of the fit (i.e. objective functions in Maerten et al., 2016) must honour the relationships between structure attitudes (and senses of shear when faults are involved) and stress orientations (Fig. 7.13). That is, σ_3 should be perpendicular to mode 1 fractures and σ_1 perpendicular to stylolithe planes

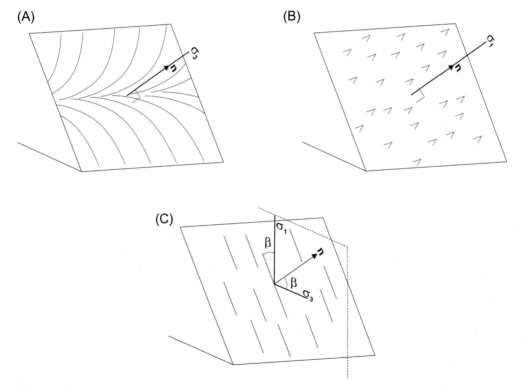

FIG. 7.13 Fitness criteria used in the geomechanical inversion for (A) mode 1 fractures, (B) stylolithic planes and (C) neoformed faults. See text for details.

but should make an angle of β with neoformed fault planes. The angle β relates directly to the angle of internal friction (see Section 3.2.3 and Eq. 3.118), which in turn is prescribed prior to modelling.

To account for the latter geometrical relationships, Maerten et al. (2016) introduced the three objective functions:

$$\text{For mode 1 fractures } C_{\text{open}} = 1 - (\hat{\sigma}_3 \hat{n})^2 \tag{7.44}$$

$$\text{For stylolithes } C_{\text{close}} = 1 - (\hat{\sigma}_1 \hat{n})^2 \tag{7.45}$$

$$\text{For neoformed faults } C_{\text{shear}} = \left| (\hat{\sigma}_3 \hat{n})^2 - \frac{1}{2}(1 + \sin\phi_i) \right| \tag{7.46}$$

where n is the normal of the considered structure plane, ϕ_i the internal angle of friction and circumflexes indicate unit vectors.

The three functions are designed to deliver values (i.e. costs after Maerten et al., 2016) comprised between 0, i.e. perfect fit, and 1, i.e. fully inconsistent result. The best fit between the simulated stress field and the measured structures is found for the minimum value of the sum of all costs.

According to Eq. (7.43), the local stress depends solely on $\alpha_1(\theta, \overline{\Phi})$, $\alpha_2(\theta, \overline{\Phi})$ and $\alpha_3(\theta, \overline{\Phi})$; thus, the expressions of the maximum and the minimum principal stresses can be rewritten as a function of these three parameters in Eqs (7.44)–(7.46). Therefore, the function to be minimised during the inversion is

$$F(\alpha_1, \alpha_2, \alpha_3) = \frac{1}{N_0 + N_c + N_s} \left(\sum_{i=1}^{N_o} C_{\text{open}}^{(i)} + \sum_{i=1}^{N_c} C_{\text{close}}^{(i)} + \sum_{i=1}^{N_s} C_{\text{shear}}^{(i)} \right) \tag{7.47}$$

with N_o, N_c and N_s are the total number of input mode 1 fractures, stylolithes and faults, respectively. The resulting total cost ranges between 0 and 1 (Fig. 7.12B).

The search for the optimal stress state is performed using a classical Monte-Carlo approach, where the explored parameters are $\alpha_1(\theta, \overline{\Phi})$, $\alpha_2(\theta, \overline{\Phi})$ and $\alpha_3(\theta, \overline{\Phi})$.

7.3.3 Concluding remarks

Paleostress reconstructions by means of geomechanical modelling present the precious advantage of predicting local stress orientations, which in case of significant perturbations may deviate markedly from those of the regional stress. In turn, the information can be used to predict fracture orientations in areas where direct observation is difficult or impossible. Thus, the methodology is very interesting when attempting to infer potential fluid flow paths in the subsurface, where major structures can be identified by geophysical methods but where less extensive secondary fractures, e.g. joints, remain well below detection thresholds.

However, the practical use of the method demands detailed field analyses (or interpretations of geophysical data) in order to sort out correctly relative chronologies and to identify unambiguously the structures coeval to faulting, which are used as the primary input in the inversions. A similar remark applies as well to the minor neoformed faults used as input as these are often arduous to discriminate from reactivated pre-existing discontinuities in the field.

A last point the user should remain aware of, concerns the objective functions used to control the fitness of the results. Eqs (7.44)–(7.46) represent stringent conditions that might not always be respected by natural structures. For example, we demonstrated the possibility of formation of mode 1 fractures non-orthogonal to the minimum principal axis of stress in Chapter 5, and stylolithic planes are not obligatory perpendicular to the maximum principal axis of stress (only stylolithe teeth are parallel to it; see Section 8.4). Admittedly, the cases involved in the objective functions of Maerten et al. (2016) are valid in a statistical sense, suggesting that a relatively high number of measurements are required to obtain robust results. Nonetheless, the approach is particularly original with respect to traditional fault slip inversion methods and definitively more evolved towards geomechanical aspects.

Further reading

Lisle, R., Orife, T., Arlegui, L., 2001. A stress inversion method requiring only fault slip sense. J. Geophys. Res. Solid Earth 106, 2281–2289.

Maerten, L., Maerten, F., Lejri, M., Gillespie, P., 2016. Geomechanical paleostress inversion using fracture data. J. Struct. Geol. 84, 197–213.

Morris, A., Ferrill, D.A., Henderson, D.B., 1996. Slip-tendency analysis and fault reactivation. Geology 24, 275–278.

Pascal, C., 2004. SORTAN, a Unix program for calculation and graphical presentation of fault slip as induced by stresses. Comput. Geosci. 30, 259–265.

Sato, K., 2006. Incorporation of incomplete fault-slip data into stress tensor inversion. Tectonophysics 421, 319–330.

One step beyond: Full determination of paleostress tensors

8.1 Paleostress magnitudes from geological objects

We already emphasised in introduction to this book that the determination of paleostresses from observation and measurement of geological objects is a traditional topic in structural geology and related fields. Stress quantification is indeed an inescapable prerequisite to constrain the rheological behaviour of rocks and for the application of mechanical theories to tectonic phenomena with some degree of confidence. The high relevance of the issue has promoted a wealth of research during the past decades in order to identify and to calibrate reliable paleostress reconstruction methods with particular focus on stress magnitudes.

A wide range of methods was subsequently established over time. Many of these involve the study of fabrics and microstructures, such as *paleostress gauges* or *paleopiezometric methods*, and stem mainly from the petrological community (e.g. Turner and Weiss, 1963). Paleopiezometric methods are methods to determine differential stresses based on the study of the characteristics of a fabric (Twiss and Moores, 2001). An example of paleopiezometer is calcite twinning, which is assumed to occur above a certain level of differential stress (see details in Chapter 6). Other microstructural/petrological methods deliver different stress quantities. Nevertheless, this class of methods is, in general, not applicable to the brittle field and will not be further discussed in the following. More complete information about paleopiezometric methods can be found in the pages 253–257 of Passchier and Trouw (2005) and in the references therein.

In this chapter, we restrict the discussion to those methods dealing with quantification of stress magnitudes, which were developed in parallel or were adapted to the paleostress reconstruction methods devised in this book, and/or involving the structures presented in Chapter 2. We will omit, during the discussion, the case of calcite twinning paleopiezometry, as it has already been addressed in much detail in Chapter 6.

8.2 Adjustment of reduced Mohr circles using failure and reactivation laws

8.2.1 Setting the problem

As extensively discussed in Chapter 4 (e.g. Section 4.1.3), inversion of fault slip data furnishes a reduced form of the stress tensor, for which information about stress magnitude is missing. Bergerat et al. (1982, 1985) proposed to introduce additional constraints to derive the six parameters of the stress tensor and, in particular, made use of failure envelopes (see Section 3.2.3). Their original attempt was later extended by Angelier (1989), who supplemented the mechanical approach with (shear) reactivation laws (see Section 3.2.2). Before discussing more in detail about these works, let us rearrange Eq. (4.4), in order to get clear views on the problem:

$$T = (\sigma_1 - \sigma_3)T_r + \sigma_3 I \tag{8.1}$$

where T is the full stress tensor expressed in the coordinate system of its eigenvectors (see Eq. 3.22), T_r is the reduced stress tensor as given in Eq. (4.5), and I is the unit matrix.

The geometrical interpretation of Eq. (8.1) in the Mohr space is straightforward (Fig. 8.1). Once T_r is known, the reconstruction of the full stress tensor requires determining (1) a scale factor, for example D, the diameter of the Mohr circle and (2) a quantity defining its position, σ_3 in Eq. (8.1), which in turn is more conveniently replaced by σ_m, the mean stress in the $(\vec{\sigma}_1, \vec{\sigma}_3)$ plane or abscissa of the centre of the Mohr circle. The two quantities are defined respectively by:

$$D = \sigma_1 - \sigma_3 \tag{8.2}$$

and

$$\sigma_m = \frac{\sigma_1 + \sigma_3}{2} \tag{8.3}$$

or

$$\sigma_m = \frac{D}{2} + \sigma_3 \tag{8.4}$$

In summary, the problem to be solved appears to be astonishingly simple. The problem of determining the two missing parameters of the stress tensor, thus stress magnitude, consists in finding the size and the position of its corresponding Mohr circle (Fig. 8.1), after having derived stress orientations and shape ratio from inversion of fault slip data. We feel intuitively that the latter is equivalent to adjusting the Mohr circle either graphically or analytically using additional independent constraints.

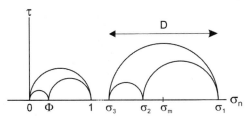

FIG. 8.1 Synthetic example showing Mohr circles corresponding respectively to the reduced stress tensor as derived from inversion of fault slip data (*left*) and to the full stress tensor (*right*). The complete restoration of the stress tensor from its reduced counterpart requires knowledge of two additional quantities. The latter gives respectively the size and the position of the "full" Mohr circle, i.e. D and σ_m in the present example (see text for details).

8.2.2 Constraints from failure envelopes

Neoformed faults measured in the field may be used to adjust the Mohr circle, as proposed in the pioneering work of Bergerat et al. (1982, 1985). The subsequent condition is that their respective poles in the Mohr space indicate the contact point between the Mohr circle and the failure envelope (Fig. 8.2A). Evidently, one needs some a priori knowledge of the failure envelope to be employed in the analysis. The latter may be determined by means of mechanical testing of relevant rock samples in the lab (Bergerat et al., 1982, 1985) or alternatively, inferred from empirical rupture laws, like, for instance, the Hoek and Brown criterion (see Section 3.2.3, Eq. 3.132) as advanced by Angelier (1989).

Once the required failure envelope is established or empirically selected, the constraint may be graphically applied. An alternative and certainly more objective approach consists in deriving the mathematical formulation of the constraint. To these aims, we need to remember that the role of the intermediate principal stress on brittle failure may be neglected (see Section 3.2.3). The mathematical derivation is therefore simplified, but the reader should not forget that the value of σ_2 is also addressed by the approach presented here. After having adjusted the Mohr circle, its value can be deduced from the shape ratio, obtained on beforehand during the inversion of the fault slip data (see Bergerat et al., 1985 for further details). In addition, the slope of the failure envelope is assumed constant (Fig. 8.2A), meaning the whole envelope mimics the modified Griffith

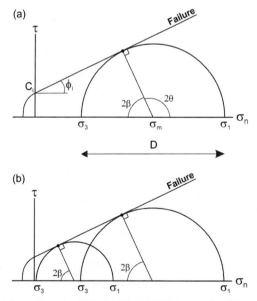

FIG. 8.2 (A) Mohr circle adjusted to a failure envelope and quantities involved in the analytical treatment presented in the text. (B) Illustration of the non-uniqueness of the solution: the depicted Mohr circles represent two out of an infinite number of Mohr circles fulfilling the failure condition. Note that σ_2 has been omitted for the sake of clarity.

criterion (Fig. 3.16D). This latter assumption is justified by the fact that the slopes of rupture curves determined in the lab for rocks show modest variations for large pressure intervals in general (e.g. Hoek and Brown, 1980).

Let us now substitute D and σ_m into Eqs (3.78) and (3.81) and replace 2θ by $180°-2\beta$ (Fig. 8.2A). After rearranging we obtain new expressions for the coordinates of the pole of the neoformed fault in the Mohr space:

$$\sigma_n = \sigma_m - \frac{D}{2}\cos 2\beta \tag{8.5}$$

$$\tau = \frac{D}{2}\sin 2\beta \tag{8.6}$$

Considering the relationship between β, the orientation of the fracture plane, and ϕ_i, the angle of internal friction, given by Eq. (3.118), Eqs (8.5) and (8.6) become:

$$\sigma_n = \sigma_m - \frac{D}{2}\sin \phi_i \tag{8.7}$$

$$\tau = \frac{D}{2}\cos \phi_i \tag{8.8}$$

By definition, the pole of a neoformed fault belongs to the failure envelope in the Mohr space. Therefore, substituting Eqs (8.7) and (8.8) in Eq. (3.117) we obtain:

$$\frac{D}{2}\cos \phi_i = C_i + \left(\sigma_m - \frac{D}{2}\sin \phi_i\right)\tan \phi_i \tag{8.9}$$

$$D = \frac{2C_i}{\cos \phi_i} + \frac{2\sigma_m \tan \phi_i}{\cos \phi_i} - D\tan^2 \phi_i \tag{8.10}$$

$$(1 + \tan^2 \phi_i)D = \frac{2}{\cos \phi_i}(C_i + \sigma_m \tan \phi_i) \tag{8.11}$$

Let us remark that:

$$(1 + \tan^2 \phi_i) = \frac{\cos^2 \phi_i}{\cos^2 \phi_i} + \frac{\sin^2 \phi_i}{\cos^2 \phi_i} = \frac{1}{\cos^2 \phi_i} \tag{8.12}$$

Combining Eqs (8.11) and (8.12) and rearranging the result, we finally obtain:

$$D = 2(\cos \phi_i C_i + \sin \phi_i \sigma_m) \tag{8.13}$$

Note that after rearranging Eq. (8.13), we find Eq. (3.120) again, which is the analytical expression of the shear failure envelope in the space of the principal stresses.

The mathematical development leads to one single equation relating the two unknowns, D and σ_m, which is obviously not sufficient to solve the problem. Eq. (8.13) simply reflects a result that was intuitively anticipated: an infinite number of Mohr circles satisfy the failure condition (Fig. 8.2B). Nevertheless, the effort is not in vain if another constraint can be added (as it will be discussed later in Section 8.2.4, Bergerat et al., 1982, 1985, estimating the vertical stress component as additional constraint).

8.2.3 Constraints from shear reactivation laws

As discussed in the previous section, the Mohr circle corresponding to the complete stress tensor is not fully constrained by the results of fault slip inversion when supplemented by the use of a rupture law only. Angelier (1989) proposed to involve reactivated fault planes in the analysis in order to bring in the missing constraint. In his approach, the Mohr circle is further adjusted based on the distribution of the poles of the reactivated faults and an inferred friction line (Fig. 8.3A).

A sufficient number of fault slip data may furnish an estimate of the slope of the friction line, when the lower boundary of the distribution of their poles is well imaged in the Mohr space. Alternatively, the slope may be inferred from empirical laws. We should note that the friction line represents initial friction here, which is the friction measured right at the start of shear motion (Byerlee, 1978), and not static friction (see Section 3.2.2). The latter assumption finds qualitative justification in the modest amount of shear displacement on the faults traditionally involved in paleostress reconstructions. As a consequence, the friction line has to pass through the origin of the Mohr space, which in turn represents a very pleasant simplification of the problem.

The graphical adjustment of the Mohr circle using a friction law is merely seldom done (Fig. 8.3A). Let us consider instead the analytical treatment of this additional constraint on the missing parameters of the stress tensor. The shear reactivation condition is met if the linear segment defining the base of the circular cap, containing the poles of the reactivated faults, coincides with the friction line. Various alternatives exist to represent this condition

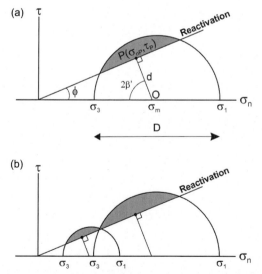

FIG. 8.3 (A) Mohr circle adjusted to a friction line and quantities involved in the analytical treatment presented in the text. (B) Illustration of the non-uniqueness of the solution: the depicted Mohr circles represent two out of an infinite number of Mohr circles fulfilling the shear reactivation condition. Grey areas indicate the circular caps containing the poles of the reactivated discontinuities. Note that σ_2 has been omitted for the sake of clarity.

mathematically. In the following, we consider the segment OP from the centre of the Mohr circle to the linear base of the cap and orthogonal to it (Fig. 8.3A). The condition is fulfilled when point P is included in the friction line, in other when the orthogonal distance between the centre of the Mohr circle and the friction line equals d. However, as there is no unique solution to the problem (Fig. 8.3B), one has to consider some normed value characteristic of all Mohr circles meeting the reactivation condition instead of the distance between O and P, which is variable from one Mohr circle solution of the problem to the other. Remarking that all Mohr circles meeting the condition are homothetic to each other (Fig. 8.3B), such a constant normed value can be, for example:

$$r = 2\frac{d}{D} \tag{8.14}$$

which is easily estimated graphically.

Keeping these preliminary remarks in mind, we now express the coordinates of point P in the Mohr space, that is we project segment OP orthogonally first on the abscissa and then on the ordinate:

$$\sigma_{nP} = \sigma_m - d\cos 2\beta' \tag{8.15}$$

$$\tau_P = d\sin 2\beta' \tag{8.16}$$

Replacing d in function of r and D (i.e. Eq. (8.14)) and β' in function of the angle of initial friction ϕ (i.e. Eq. 3.111), we find:

$$\sigma_{nP} = \sigma_m - r\frac{D}{2}\sin\phi \tag{8.17}$$

$$\tau_P = r\frac{D}{2}\cos\phi \tag{8.18}$$

By definition (see Section 3.2.2):

$$\tan\phi = \frac{\tau_P}{\sigma_{nP}} \tag{8.19}$$

Combining Eqs (8.17) and (8.19), we find:

$$\tan\phi = \frac{r\dfrac{D}{2}\cos\phi}{\sigma_m - r\dfrac{D}{2}\sin\phi} \tag{8.20}$$

$$\left(\sigma_m - r\frac{D}{2}\sin\phi\right)\frac{\sin\phi}{\cos\phi} = r\frac{D}{2}\cos\phi \tag{8.21}$$

and finally,

$$D = 2\frac{\sin\phi}{r}\sigma_m \tag{8.22}$$

We remark that setting $r = 1$, that is $d = D/2$ in Eq. (8.22) leads to an equation similar to Eq. (8.13), where $\phi = \phi_i$ and $C_i = 0$.

As in the case of failure in shear devised in Section 8.2.2, Eq. (8.22) gives a relationship between the two unknowns, D and σ_m, and expectedly does not fully constrain the problem on its own.

8.2.4 Combination of failure and shear reactivation laws and final remarks

The two mathematical developments, presented in Sections 8.2.2 and 8.2.3, lead to two independent equations, (8.13) and (8.22), relating the diameter of the Mohr circle (i.e. its scale) to the coordinates of its centre (i.e. its position), and thus provide a unique solution to the problem (Fig. 8.4), as demonstrated by Angelier (1989). The derivation of the magnitudes of the three principal stresses from D, σ_m and Φ is seldom done.

Obviously, any one of the two relationships may be used to determine the full stress tensor, provided that an additional constraint is brought in. For example, Bergerat et al. (1982, 1985) estimated the respective paleodepths of their study sites (i.e. the depths at which faulting occurred) and derived the corresponding lithostatic pressures, assuming rock densities. At each site, the calculated pressure was taken equal to the magnitude of the principal axis of stress, which had been found vertical (or nearly vertical) after inversion of the fault slip data.

As such, the procedure was permitted to locate one of the principal stresses, that is the one whose corresponding axis was found (nearly) vertical, on the abscissa in the Mohr space. The relative positions of the two other principal stresses were deduced from the value of the shape ratio, and finally, the diameter of the Mohr circle was varied until the circle became tangential to the failure envelope.

In mathematical terms, the approach is equivalent to establishing the equations relating the vertical principal stress to the two unknowns of the Mohr circle, D and σ_m, for normal, strike-slip and reverse stress regimes. The derivation of these three equations from Eqs (3.38), (8.3) and (8.4) is straightforward and results in:

$$\sigma_V = \sigma_m + \frac{D}{2} \text{ for } \sigma_1 = \sigma_V \tag{8.23}$$

$$\sigma_V = \sigma_m + \left(\Phi - \frac{1}{2} \right) D \text{ for } \sigma_2 = \sigma_V \tag{8.24}$$

and

$$\sigma_V = \sigma_m - \frac{D}{2} \text{ for } \sigma_3 = \sigma_V \tag{8.25}$$

Finally, selecting the relevant equation amongst the three given above and combining it with Eq. (8.13) or (8.22) permits to calculate D and σ_m and, thus, to restore the six parameters of the paleostress tensor.

We should note, however, that the use of lithostatic pressure to complete the analysis is limited to sites where paleodepths can be estimated independently, a condition that is often fulfilled with significant uncertainty or almost impossible to meet in, for example

(a) Adjustment to the failure envelope

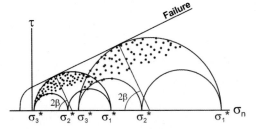

(b) Adjustment to the friction line

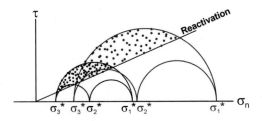

(c) Adjustment combining failure and reactivation

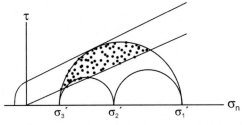

FIG. 8.4 Illustration of the 'Bergerat–Angelier approach' to determine the full paleostress tensor. Inversion of fault slip data delivers the orientations of the principal axes of stress and the shape ratio (i.e. relative position of the intermediate principal stress in the Mohr space), which is an unscaled Mohr circle of unknown location. Adjustment of the Mohr circle to the failure envelope (A) or to the friction line (B) results in an infinite number of solutions, but combination of the two criteria (C) allows for adjusting a unique Mohr circle (see mathematical demonstrations in Sections 8.2.2 and 8.2.3) and, thus, for finding a unique full stress tensor. Note, however, that in the absence of independent quantification of pore paleopressure the methodology returns effective stresses (indicated with primes in (C)). Asterisks in (B) and (C) recall that the obtained stress values remain unconstrained.

basement areas. Stress states with no vertical axis (i.e. non-Andersonian stress states) represent obviously a second limitation to the approach, though less common than the previous one.

It appears somewhat surprising that, although established more than three decades ago, the 'Bergerat–Angelier approach' (Fig. 8.4) has been rarely employed or further explored. An exception to this statement is the tentative extension of the method to mode 1 fractures by

André et al. (2001). This latter approach proposes a very interesting combination of the use of failure laws, as in Angelier (1989), with Jolly and Sanderson's theory (see Chapter 5). To the opinion of the author, the method of André et al. (2001) presents, however, serious inconsistencies from the mechanical point of view. A second exception is the application of the 'Bergerat–Angelier approach' by Choi et al. (2013) to their paleostress study in southeast Korea. This study maintained the combination of failure and friction laws but as a slight modification of the original approach, the friction line was quantified according to the relationship between orientations of T-fractures (Fig. 2.10) and fault friction, as previously established by Choi et al. (1996).

Admittedly, the inherent limitations of the 'Bergerat–Angelier approach' might explain why the scientific community has been so timorous with it until now. The need for detecting and measuring neoformed faults, a prerequisite that is not systematically warranted in the field, precludes the use of the failure condition from scratch.

Even when data from neoformed faults are collected, the choice of the failure law, either empirical or driven by experimental work on relevant rock samples, represents a major risk for the analysis. An incorrect choice of the failure envelope might result in significant errors in stress magnitudes and might explain the surprisingly high magnitudes found by Bergerat et al. (1982, 1985). Incorrect selection of the friction line might also lead to serious discrepancies, though the line can potentially be constrained with sufficient accuracy by the fault slip data, as suggested in Fig. 8.4. In brief, the method represents a step forward for paleostress reconstruction, but its outcomes need to be regarded with circumspect eye.

As a final remark, we should note that the 'Bergerat–Angelier approach' delivers, obviously, the magnitudes of the stresses responsible for the observed ruptures and slips, in other words the magnitudes of the effective stresses (see Section 3.2.4 for definition). Therefore, these need to be corrected for pore fluid paleopressure, in order to get proper estimates of tectonic stress magnitudes.

8.3 Pore fluid paleopressure determinations from analyses of fluid inclusions

Minimum pore paleopressure values may be obtained assuming *hydrostatic pressure*, which is pore pressure corresponding to the weight of the fluid column from the Earth surface to the depth level under scope (*lithostatic pressure* corresponding to the weight of the rock column). Evidently, the approach may be adopted only if paleodepths can be safely estimated from independent geological evidence (e.g. Bergerat et al., 1985; Angelier, 1989). Another certainly more serious drawback of the assumption is that hydrostatic conditions require hydraulic connection between underground fluids and the Earth surface. Such a situation is plausible, in particular, at relatively shallow depths, provided no impermeable rock (e.g. shale layers) disrupts hydraulic connection with the surface. However, it is often difficult to unveil the past hydrological conditions from field observations alone, and present-day measurements in boreholes show that fluid pressures above hydrostatic values are indeed common, especially in sedimentary basins (see chapter 2 in Zoback, 2007).

Lespinasse and Cathelineau (1995) and Lespinasse (1999) proposed to combine paleostress determination with analyses of fluid inclusions in order to overcome these limitations. In the following, we address the main aspects concerning fluid inclusion analyses; the reader is invited to consult the classical book of Roedder (1984) for an exhaustive presentation. *Fluid inclusions (FI)* are minute quantities of fluids trapped in microscopic cavities within crystals (Fig. 8.5; Roedder, 1984). Fluid inclusions are found in a broad range of minerals but are more easily studied in quartz. They usually host multiphase fluids, where typically H_2O and CO_2 are the dominant phases (Fig. 8.5A) and salts, CH_4 and N_2 the less abundant ones. They can be either *primary fluid inclusions* formed during mineral growth and, thus, representing the composition of the original fluid, or *secondary fluid inclusions* resulting from later fracturing of the minerals and, therefore, displaying the characteristics of the fluids infiltrated during the corresponding deformation event. Secondary FI can be isolated, as primary ones commonly are, but form very often planar *inclusion trails* or *fluid inclusion planes* (FIPs, Fig. 8.5B). FIPs represent healed microfractures where each individual FI is a remaining cavity that survived the healing process (Tuttle, 1949). Comparison with mesoscale structures shows that these microfractures initiated in general as mode 1 fractures (Wise, 1964; Lespinasse and Pêcher, 1986; Laubach, 1989; Lespinasse and Cathelineau, 1995).

In practice, FI are studied on doubly polished thick sections with the aim of determining composition and density of the fluids they host. Various sophisticated techniques (see pages 286–287 in Passchier and Trouw, 2005 for a comprehensive summary) may be employed in this respect, but thermal analysis of the FI with the help of a heating–freezing stage and under a standard optical microscope remains the most common and cheap approach. The microthermometric analysis is carried out on one FI at a time. It consists in measuring temperatures corresponding to visible modifications of the phases it contains, for example melting or homogenisation temperatures (Roedder, 1984), the operator controlling the evolution of the FI during heating or freezing under the microscope.

The microthermometric data reveal the composition of the FI that, in turn, is used to derive density. Finally, the density defines the isochore of the fluid, meaning its pressure–temperature path at constant volume. Independent knowledge of paleodepth and paleo-geothermal gradient (Lespinasse and Cathelineau, 1995) or constraints from relevant geothermometers (Jaques and Pascal, 2017 and Fig. 8.6) allows for determining the most likely pressure range related to the emplacement of the fluid, thus the pore fluid paleopressure.

To conclude, we should note that different strategies were adopted to include FI analyses in paleostress computations in the past. For instance, Lespinasse and Cathelineau (1995) combined the 'Bergerat–Angelier approach' with analyses of FIPs. Synchronicity between the studied FIPs and the collection of strike-slip faults they measured in the field

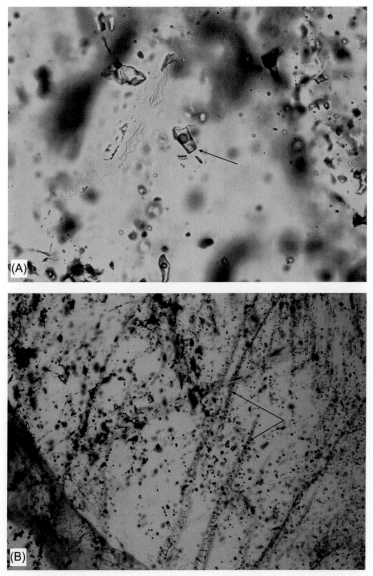

FIG. 8.5 Examples of fluid inclusions in quartz samples collected in the mines of Panasqueira, Portugal (photos courtesy of Luís Jaques). (A) Primary fluid inclusion with prominent CO_2 bubble (scale bar of 10 μm). (B) Fluid inclusion trails or FIPs (scale bar of 100 μm).

was the key assumption of their work. The assumption was supported by the observation that the orientation of the minimum principal stress, found after fault slip data inversion, was perpendicular to the FIPs on average, the latter being viewed as mode 1 microcracks. Alternatively, André et al. (2001) and, more recently, Jaques and Pascal (2017) studied FI in quartz veins in order to integrate their respective results with inversion of tensile fractures

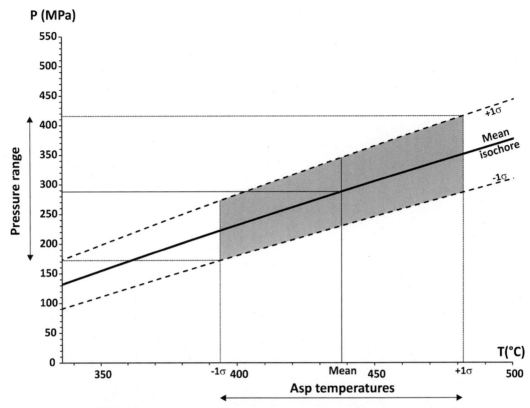

FIG. 8.6 Pore fluid paleopressure constrained by combination of fluid inclusion analyses and the use of a geothermometer (arsenopyrite, i.e. Asp, in the present case). The mean isochore was calculated from 16 isochores, themselves derived from analyses of ~800 fluid inclusions. Arsenopyrite crystallisation temperatures were determined from geochemical analyses on 10 samples (~400 punctual electron microprobe analyses), where σ represents standard deviation. *Data from Jaques and Pascal (2017).*

(see Chapter 5) and modified versions of the 'Bergerat–Angelier approach'. In both studies, the targeted FI were primary and supposedly formed shortly after rupture of the tensile fractures. In summary, the use of FI to determine paleopressures is particularly advantageous and presents enough flexibility to complement the different kinds of paleostress analyses at hand.

8.4 Stylolithe paleopiezometry

8.4.1 Self-affinity of stylolithe surfaces

Stylolithes are pressure-solution structures developed perpendicularly to the axis of maximum principal stress and mainly found in carbonate rocks in the brittle field of deformation (Park and Schot, 1968 and Section 2.2.4). Stylolithes show, in general, pronounced

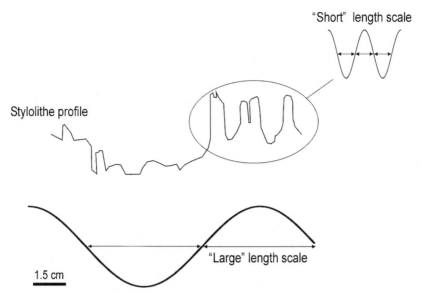

FIG. 8.7 Typical 1D topographic profile of a stylolithe depicting its roughness at both "large" and "short" length scales or, alternatively, wavelengths. The profile corresponds to the digitised left half part of the stylolithe shown previously in Fig. 2.6B.

surface roughness[a] (Fig. 2.6), whether they are observed at relatively large or relatively short length scales (Fig. 8.7).

Major advances in stylolithe research, during the past 2 decades, owed to the improvement of high-resolution mechanical or laser profilometers, which allowed for measuring deviations down to 10s of micrometres of small sample surfaces (see review in Toussaint et al., 2018). The measurements demonstrated that *self-affinity* is one of the most remarkable properties of stylolithe surface roughness (Renard et al., 2004; Schmittbuhl et al., 2004).

Self-affinity defines the property of some objects to remain invariant by anisotropic affine transformation (Mandelbrot, 1985). For example, in the case of a self-affine rough surface, the surface obeys the affine transformation:

$$\Delta x \rightarrow b\Delta x \quad \Delta y \rightarrow b\Delta y \quad \Delta z \rightarrow b^{H}\Delta z \tag{8.26}$$

where Δx, Δy and Δz are amplitudes in the x, y and z directions, b is a real number, the transformation factor and H is the *Hurst or roughness exponent*. The Hurst exponent varies between 0 and 1 and its highest values characterise relatively smooth surfaces. When H is equal to 1 the surface is self-similar.

---

[a]The term is self-explanatory, especially when considering the opposite notion: surface smoothness. A more elaborated definition of surface roughness involves quantification of the deviations of the normal vector, when this one is moved over the surface under scope, with respect to an idealised form of the surface (i.e. a plane in case of stylolithes).

In summary, roughness is a statistical property of surfaces, which can be quantified by the Hurst exponent. Various signal processing methods exist to calculate roughness from topographic profiles of surfaces (see Candela et al., 2009). Amongst these, the *Fourier power spectrum* method is one of the most employed methods to treat not only stylolithes (Renard et al., 2004) but also fracture surfaces that are either man-made (Schmittbuhl et al., 1995) or of tectonic origin, like faults (Candela et al., 2009) or joints (Nigon et al., 2017).

The Fourier power spectrum, $P(k)$, is defined as the square of the modulus of the Fourier transform of the studied signal expressed in function of the wave number:

$$k = \frac{2\pi}{\lambda} \qquad (8.27)$$

where λ is wavelength.

In short, the Fourier transform is a classical mathematical tool used to break down a signal in its different wavelength components, and the Fourier power spectrum indicates how much signal exists for a given wavelength. The interesting advantage of the Fourier power spectrum is that for a self-affine signal (see Schmittbuhl et al., 1995 and references therein for further details):

$$P(k) \sim k^{-2H-Dim} \qquad (8.28)$$

where $Dim = 1$ for a 1D profile and H is the Hurst exponent as introduced above.

In other words, the Fourier power spectrum of a self-affine 1D profile draws a straight line on a log-log plot. Its slope is proportional to the Hurst exponent and ordinate values are proportional to surface roughness amplitudes at their corresponding wave-numbers.

In the case of stylolithes (Fig. 8.8), the Fourier power spectrum analysis, or the application of a similar processing method, reveals indeed two distinct self-affine regimes (Renard et al., 2004; Schmittbuhl et al., 2004; Ebner et al., 2009, 2010; Rolland et al., 2012, 2014; Beaudoin et al., 2016): one at large length scales (or wavelengths) and associated with $H \sim 1$, and another at short length scales with $H \sim 0.5$, the two regimes being separated by a well-defined crossover length, L_c, of about 1 mm typically.

8.4.2 Chemico-mechanical model for stylolithe formation (Schmittbuhl et al., 2004)

The existence of two self-affine scaling regimes for stylolithe roughness is highly suggestive of different stylolitisation mechanisms acting at distinct length scales. Based on measurements of Renard et al. (2004), Schmittbuhl et al. (2004) proposed a chemico-mechanical model to take into account the peculiar roughness behaviour of stylolithes (see also Rolland et al., 2012 for detailed mathematical treatment of the problem).

Stylolitisation is assumed to occur at the fluid–rock interface of a pre-existing discontinuity in the rock (e.g. a bedding surface), whose mean plane is perpendicular to the maximum principal stress (Fig. 8.9). Note that only one wall of the discontinuity, either the upper or the lower one, is considered in the mathematical development of the model. The fluid–rock interface has to remain all along the process and thus, pore fluid pressure

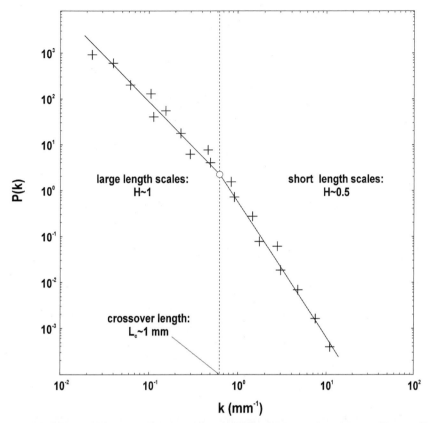

FIG. 8.8 Synthetic log-log graph depicting the two distinct self-affine regimes of stylolithe roughness. *P(k)* is Fourier power transform and *k*, wave number (see text for details). *H* and L_c indicate the Hurst exponent and crossover length. *Crosses* represent virtual data points.

is supposed to be equal to the magnitude of the stress. Nevertheless, the fluid must be drained. In the opposite case, saturation of the fluid with soluble species would be quickly reached and stylolitisation would be consequently stopped.

The model involves a soluble matrix with randomly distributed less soluble (or potentially insoluble) grains, mimicking, for example, carbonate rocks containing small amounts of randomly distributed mica grains. At the stylolithe surface, less soluble grains resist dissolution and create a roughening destabilising force by pinning the surface of the discontinuity and promoting growth of peaks and teeth. This destabilising force is counteracted by two smoothening stabilising forces, which tend to flatten the relief of the stylolithe surface. The first one is surface tension, acting at short length scales, and the second one is the elastic reaction of the material operating at large length scales (note that elastic forces can also be destabilising in some circumstances, Rolland et al., 2012). Thus, depending on scale, one or the other stabilising force intervenes in the process of

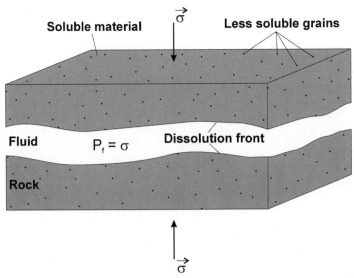

FIG. 8.9 Chemico-mechanical model of stylolithe formation according to Schmittbuhl et al. (2004) and Rolland et al. (2012). Stylolitisation occurs at the fluid–rock interface of a discontinuity whose mean plane is perpendicular to the applied maximum principal stress and pore fluid pressure, P_f, is equal to σ. Less soluble grains create a destabilising roughening force by pinning the stylolithe surface, which is in turn resisted by two types of stabilizing smoothening forces, long-range elastic forces and short-range surface tension (see details in text). Note that only one wall of the discontinuity is considered in the development of the mathematical model. Other stress components are omitted for the sake of clarity.

stylolitisation, and this duality explains the two distinct self-affine scaling regimes evidenced for stylolithe surfaces (Fig. 8.8). Therefore, the crossover length marks the boundary between the respective domains of action of the two stabilising forces.

Schmittbuhl et al. (2004) demonstrated that the value of the crossover length was dependent on the magnitude of the applied mean stress and derived the following equation:

$$L_c = \frac{\gamma E}{\beta p D} \tag{8.29}$$

where γ is surface energy of the solid–fluid interface, E is Young's modulus of the solid, p is mean stress as given in Eq. (3.34), D is differential stress as in Eq. (8.4) and b is a dimensionless number in function of the Poisson's ratio, ν, such as

$$\beta = \frac{\nu(1-2\nu)}{\pi} \tag{8.30}$$

In the common case of compaction stylolithes (i.e. sedimentary stylolithes created by lithostatic load), crossover length remains constant independently of the orientation of the topographic profile, which is measured parallel to the mean stylolithe surface (Renard et al., 2004; Schmittbuhl et al., 2004; Ebner et al., 2009; Rolland et al., 2014).

The latter observation implies an isotropic stress state parallel to the stylolithe plane, which obviously means parallel to the horizontal plane for compaction stylolithes. Consequently, one may assume that the two horizontal principal stresses, σ_H and σ_h, are equal. In addition, remarking that compaction stylolithes are mostly generated in situations of tectonic quiescence, where lithostatic loads are supposed to dominate, one may assume that the magnitude of the vertical principal stress, σ_V, is significantly higher than that of the two horizontal principal stresses, which in turn result of the elastic response (i.e. Poisson's effect) of the rock mass solely. Finally, the rock mass being laterally confined, horizontal strain is zero.

Taking into account the previous considerations, replacing σ_1, σ_2 and σ_3 by σ_V, σ_H and σ_h, respectively, and combining Eqs (3.90) and (3.92), we find the classical equation:

$$\sigma_H = \sigma_h = \frac{v}{(1-v)}\sigma_V \tag{8.31}$$

The mean and differential stresses may be respectively re-written:

$$p = \frac{1}{3}(\sigma_V + 2\sigma_H) \tag{8.32}$$

$$D = \sigma_V - \sigma_H \tag{8.33}$$

Finally, substituting Eqs (8.32) and (8.33) into Eq. (8.29), the vertical principal stress is:

$$\sigma_V = \sqrt{\frac{\gamma E}{\alpha \beta L_c}} \tag{8.34}$$

where

$$\alpha = \frac{1}{3}\frac{(1+v)(1-2v)}{(1-v)^2} \tag{8.35}$$

In summary, analysis of compaction stylolithe roughness permits to determine the crossover length, L_c, and Eq. (8.34) furnishes an estimate of the vertical principal stress (i.e. σ_1), which in turn is used in Eq. (8.31) to compute the magnitude of the horizontal stress components (i.e. σ_2 and σ_3).

The above development is particularly useful when trying to estimate paleodepths by assuming or measuring the density of the considered rock column (Ebner et al., 2009). The observation that L_c varied with depth in a sedimentary sequence on a cliff of the Cirque de Navacelles, France, demonstrated that this length scale could be used to estimate paleodepth. Noteworthy, the maximum paleodepth the method is capable to deliver corresponds to the depth the stylolitisation process shut down, the analysis remains blind to the potential occurrence of further burial. At first view, stylolithe paleopiezometry appears nevertheless to be rather limited when tectonic stresses are the primary focus of research.

Ebner et al. (2010) extended the original principles of stylolithe paleopiezometry to tectonic stylolithes. Their key finding, later confirmed by Beaudoin et al. (2016), was on the azimuthal variability of crossover length with respect to the orientation of the topographic

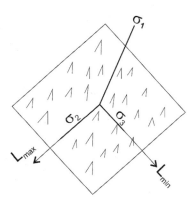

FIG. 8.10 Geometrical relationships between orientations of the principal axes of stress and maximum, L_{max}, and minimum, L_{min}, crossover lengths, respectively, for tectonic stylolithes.

profiles analysed parallel to the stylolithe plane. Crossover length was found to vary gradually from a maximum value in one direction to a minimum value in the perpendicular direction. Examination of Eq. (8.29) shows that crossover length is inversely proportional to differential stress. Ebner et al. (2010) assumed, therefore, that maximum and minimum crossover lengths corresponded to the minimum and the maximum differential stresses allowed for a plane perpendicular to the stylolithe, i.e. $\sigma_1 - \sigma_2$ and $\sigma_1 - \sigma_3$, respectively.

In other words, tectonic stylolithes present maximum crossover length parallel to the $(\vec{\sigma}_1, \vec{\sigma}_2)$ plane and minimum crossover length parallel to the $(\vec{\sigma}_1, \vec{\sigma}_3)$ plane (Fig. 8.10), the quantity taking intermediate values for all other planes perpendicular to the stylolithe surface. A major implication of the assumption is that the 3D analysis of stylolithe roughness furnishes directly the orientations of the three principal axes of stress, σ_2 and σ_3 being parallel to the directions of maximum and minimum crossover lengths, respectively, and σ_1 perpendicular to the stylolithe plane, as expected.

Ebner et al. (2010) attempted, moreover, to estimate stress magnitudes from roughness analyses of tectonic stylolithes. The most audacious step in their approach resides in the assumption that Eq. (8.29) can be applied, independently, for the two directions of the stylolithe surface corresponding, respectively, to the maximum and minimum crossover length. The application of this 2D equation to compaction stylolithes is fully justified by the axial symmetry of the problem. In the present case, where the stress is anisotropic for all sections of the 3D space, the solution is not satisfactory from a theoretical perspective, but represents, however, the only one available to date. The reader should remain aware of this potential limitation.

Following Ebner et al. (2010), we restrict the analysis to the case of vertical stylolithes and assume strike-slip stress regime. Let us use the notation for the principal axes of stress already introduced above for compaction stylolithes and write: $\sigma_1 = \sigma_H$, $\sigma_2 = \sigma_V$ and $\sigma_3 = \sigma_h$. The goal of the mathematical development is to derive two equations expressing the two horizontal stress components in function of the vertical one. The two equations will estimate the magnitudes of the horizontal stresses, provided that the magnitude of the vertical stress is determined independently. Note that the following development is easily adapted to reverse stress regimes.

We define the 2D differential stresses according to the $(\vec{\sigma}_H, \vec{\sigma}_V)$ plane and the $(\vec{\sigma}_H, \vec{\sigma}_h)$ plane, respectively,

$$D_V = \sigma_H - \sigma_V \tag{8.36}$$

$$D_h = \sigma_H - \sigma_h \tag{8.37}$$

Let us rearrange Eq. (8.29) and apply it to the two directions defined by the maximum, L_V, and the minimum, L_h, crossover lengths, respectively. We find:

$$p = \frac{\gamma E}{\beta L_V D_V} \tag{8.38}$$

and

$$p = \frac{\gamma E}{\beta L_h D_h} \tag{8.39}$$

Combining Eqs (8.38) and (8.39), we obtain:

$$L_V D_V = L_h D_h \tag{8.40}$$

Let us substitute Eqs (8.36) and (8.37) in Eq. (8.40) and derive the following relationship among the magnitudes of the three principal stresses:

$$\sigma_h = \sigma_H - \frac{L_V}{L_h}(\sigma_H - \sigma_V) \tag{8.41}$$

We have derived an equation to estimate σ_h. Now, let us find a second equation for σ_H. For the sake of commodity, we introduce the constant:

$$a = \frac{\gamma E}{\beta} \tag{8.42}$$

Eq. (8.39) can thus be rewritten:

$$pD_h = \frac{a}{L_h} \tag{8.43}$$

We write the mean stress, p, and the differential stress, D_h, in function of the magnitudes of the three principal axes of stress (i.e. Eqs 3.34 and 8.37, respectively) and Eq. (8.43) becomes:

$$\frac{1}{3}(\sigma_H + \sigma_h + \sigma_V)(\sigma_H - \sigma_h) = \frac{a}{L_h} \tag{8.44}$$

Combining Eqs (8.41) and (8.44), we find:

$$\left[\left(2 - \frac{L_V}{L_h}\right)\sigma_H + \left(1 + \frac{L_V}{L_h}\right)\sigma_V\right]\frac{L_V}{L_h}(\sigma_H - \sigma_V) = \frac{3a}{L_h} \tag{8.45}$$

Let us rearrange Eq. (8.45) in order to derive a quadratic equation for σ_H:

$$\frac{L_V}{L_h}\left(2-\frac{L_V}{L_h}\right)\sigma_H^2 + \frac{L_V}{L_h}\left(2\frac{L_V}{L_h}-1\right)\sigma_V\sigma_H - \frac{L_V}{L_h}\left(1+\frac{L_V}{L_h}\right)\sigma_V^2 - \frac{3a}{L_h}=0 \tag{8.46}$$

Eq. (8.46) leads to two solutions:

$$\sigma_H = \frac{-\frac{L_V}{L_h}\left(2\frac{L_V}{L_h}-1\right)\sigma_V \pm \sqrt{\left(3\frac{L_V}{L_h}\sigma_V\right)^2 + 12\frac{L_V}{L_h}\left(2-\frac{L_V}{L_h}\right)\frac{a}{L_h}}}{2\frac{L_V}{L_h}\left(2-\frac{L_V}{L_h}\right)} \tag{8.47}$$

Based on their data, Ebner et al. (2010) showed empirically that the two solutions of Eq. (8.46) are two real numbers with different signs. The one corresponding to the physical solution is evidently the positive root of the quadratic equation.

In summary, the method requires an independent estimation of σ_V. The latter value and the two determined crossover lengths in Eq. (8.47) furnish an estimate of σ_H, which in turn is employed to derive σ_h from Eq. (8.41). The reader should note that, in virtue of the background hypothesis of a thin water film pressurised to the level of the normal stress applied to the stylolithe surface (Fig. 8.9), the computed stresses are the tectonic stresses and not the effective stresses.

Before closing the topic of stylolithe paleopiezometry, let us discuss the sensitivity of the results with respect to the input parameters intervening in Eq. (8.29). Variations in surface energy and Poisson's ratio, in their respective reasonable ranges, have a rather limited impact of less than 10% on the obtained stress values (Beaudoin et al., 2016). The influence of the Young's modulus is far more critical and can lead to significant uncertainties (Rolland et al., 2012).

The Young's modulus can be measured on selected samples or deduced from seismic studies, however the mechanical properties of the present-day rock might differ drastically from those at the time the stylolithes formed. For instance, in their paleopiezometric study, Ebner et al. (2009) derived Young's moduli both from seismic wave velocity measurements, in the lab, and by means of computations involving the different paleodepths of the studied stylolithes. The two methods returned values differing by up to ∼80 GPa, and the authors concluded that the second method was more reliable to account for the mechanical conditions that assisted stylolitisation. That the sediments might have been more porous than present-day or even not lithified yet are obvious reasons to explain the mismatch. In addition, one cannot exclude that the mechanical properties of the fluid–rock interface differed from those of the rock mass during stylolitisation.

In spite of this latter difficulty, it is however interesting to note that the recent results of Beaudoin et al. (2016) seem to support the validity of stylolithe paleopiezometry methods. Beaudoin et al. conducted paleostress reconstructions based on both stylolithe and calcite twinning paleopiezometry and showed that the differential stresses returned by the two methods were in good agreement.

Further reading

Angelier, J., 1989. From orientation to magnitudes in paleostress determinations using fault slip data. J. Struct. Geol. 11, 37–50.

Beaudoin, N., Lacombe, O., 2018. Recent and future trends in paleopiezometry in the diagenetic domain: insights into the tectonic paleostress and burial depth history of fold-and-thrust belts and sedimentary basins. J. Struct. Geol. 114, 357–365.

Bergerat, F., Bergues, J., Geyssant, J., 1985. Estimation des contraintes liées à la formation de décrochements dans la plateforme d'Europe du Nord. Geol. Rundsch. 74, 311–320.

Lespinasse, M., Cathelineau, M., 1995. Paleostress magnitudes determination by using fault slip and fluid inclusions planes data. J. Geophys. Res. Solid Earth 100, 3895–3904.

Rolland, A., Toussaint, R., Baud, P., Schmittbuhl, J., Conil, N., Koehn, D., Renard, F., Gratier, J.-P., 2012. Modeling the growth of stylolites in sedimentary rocks. J. Geophys. Res. Solid Earth 117, B06403.

Toussaint, R., Aharonov, E., Koehn, D., Gratier, J.-P., Ebner, M., Baud, P., Rolland, A., Renard, F., 2018. Stylolites: a review. J. Struct. Geol. 114, 163–195.

9

Some examples of applications
of stress inversion methods
in tectonic analyses

9.1 Fault slip data, paleostresses and evolution of the Alpine–Mediterranean system

The first example is taken from the classical study of the paleostresses related to the evolution of the Cenozoic Alpine–Mediterranean system published by Bergerat (1987) and Le Pichon et al. (1988). To note, this work was conducted in the early years of the development and application of paleostress determination methods.

The study targeted primarily on the measurement of mesoscale fractures in rocks weakly affected by Alpine folding in Western–Central Europe. About 15,000 fracture measurements were carried out by Bergerat and co-workers, the vast majority of them were collected in France and Germany, and the details were published in various papers (see references in Bergerat, 1987). Special attention was paid to ~NW-SE prominent faults in Germany, namely the Pfahl and Franconian lineaments, and to the ~N-S precursor lineaments of the west European Rift (Upper Rhine and Saône grabens). The fault slip inversion methods of Angelier (1979, 1984) and Angelier et al. (1982) were employed to determine paleostress tensors. Heterogeneous data were typically collected at the measurement sites, requiring careful data separation into homogeneous sets (see Section 4.2.6). Relative chronologies were established at many sites, but determination of absolute chronologies was restricted to areas where convenient stratigraphical record was exposed (e.g. in southern France, Bergerat, 1987). Consequently, most age attributions relied on relative chronologies and correlations between tensors with similar stress regimes and orientations.

The results were supplemented by additional paleostress determinations in Portugal (Lepvrier and Mougenot, 1984), Spain (Guimerá, 1983, 1984), northern Africa and Arabia (Letouzey and Trémolières, 1980). In order to compare the results of the paleostress analysis with the relative motions between Africa and Eurasia in Cenozoic times, the results of the plate-scale kinematic reconstructions of Klitgord and Schouten (1986), based on Atlantic magnetic anomalies, and of Savostin et al. (1986), based on Arctic magnetic anomalies, were considered. Four major tectonic phases or stress events were identified: (1) Late Eocene N-S compression, (2) Oligocene E-W extension, (3) Early Miocene NE-SW compression and (4) Late to post Miocene NW-SE compression. In the following,

Paleostress Inversion Techniques. https://doi.org/10.1016/B978-0-12-811910-5.00011-7

we summarise supporting field observations and major outcomes for each tectonic phase. Fig. 9.1 presents the results obtained for the Late Eocene and Late Miocene phases, in particular.

9.1.1 Late Eocene ($\sim$40 Ma) N-S compression

The paleostress study of Bergerat (1987) evidenced a remarkably constant N-S compression in (present-day) Western–Central Europe during Late Eocene (Fig. 9.1A) with azimuths for the reconstructed maximum principal stresses comprised between N170°E and N010°E. Most reconstructed paleostress tensors showed strike-slip regime with σ_3 horizontal. The age of the tectonic phase was constrained, in particular, on outcrops located in southern France (at the Luberon anticline in northern Provence). Noteworthy, both the age and the stress orientations of this tectonic phase are coherent with the Pyrenean orogeny. In agreement with the stress orientation, dextral strike-slip took place along the $\sim$NW-SE Franconian and Pfahl lineaments, while the precursor lineaments of the western parts of the European Cenozoic Rift System (Upper Rhine graben) were accomodating sinistral displacements.

Both kinematic reconstructions of Klitgord and Schouten (1986) and Savostin et al. (1986) located the rotation pole of Africa with respect to Eurasia, northwest of present-day Portugal, for the period of time from 54 to 35 Ma (Fig. 9.1A). Thus, the two analyses predicted N-S motion vectors between the two continents in excellent agreement with the paleostress pattern restored by Bergerat (1987).

9.1.2 Oligocene ($\sim$33 Ma) E-W extension

The stress regime switched from strike-slip in Eocene to normal in Oligocene. Interestingly, this change in stress regime occurred by simple permutation between σ_1 and σ_2, σ_3 remaining horizontal and oriented E-W (fig. 11 in Bergerat, 1987). As in Eocene times, stress orientations in the European platform were remarkably uniform with minimum principal stress axes varying in azimuth between N075°E and N105°E.

The Oligocene E-W extension promoted the opening of the European Cenozoic Rift System, which developed in series of grabens from the North Sea (i.e. Lower Rhine graben) to southern France (i.e. Provence graben) and from Czech Republic (i.e. Eger graben) to the French Massif Central (i.e. Limagne graben). The main grabens accumulated thousands of metres of sediments.

For the period of time from 35 to 20 Ma, Savostin et al. (1986) computed northwestward displacement of Africa with respect to Eurasia, whereas Klitgord and Schouten (1986) concluded on motion due north. Le Pichon et al. (1988) speculated that the discrepancy between both kinematic reconstructions was caused by westward shift of the European area located west of the European Cenozoic Rift System with respect to the rest of Eurasia.

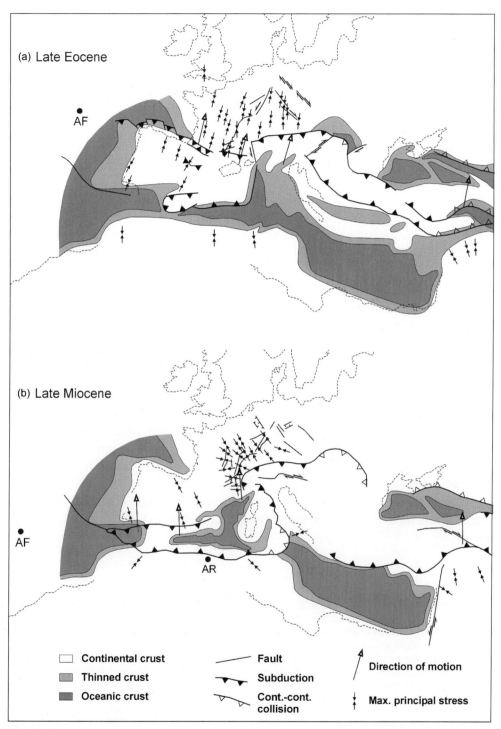

(a) Late Eocene

AF

(b) Late Miocene

AF

AR

Continental crust   ——— Fault   Direction of motion

Thinned crust   ▼——— Subduction

Oceanic crust   ▽——— Cont.-cont.
collision   Max. principal stress

FIG. 9.1 Caption on next page.

9.1.3 Early Miocene (22–20 Ma) NE-SW compression

In Early Miocene, the stress state changed drastically to NE-SW compression with rather constant stress orientations in Western–Central Europe (fig. 15 in Bergerat, 1987). As a result, the Franconian and Pfahl lines were reactivated as thrust faults and the border faults of the Upper Rhine graben experienced dextral strike-slip.

The age of the Early Miocene phase is particularly well constrained in south-eastern France, where Chattian limestones were affected by NE-SW compression in contrast to the unconformably overlying Burdigalian molasse. Thus, the estimated age for the NE-SW compression was Aquitanian (i.e. ~23 to ~20 Ma) at the location of the observation.

Similar to the kinematic results related to Oligocene times, the solutions of Savostin et al. (1986) and Klitgord and Schouten (1986) showed significant discrepancy: NNE displacement of Africa relative to Eurasia for Savostin et al. (1986), but NW motion for Klitgord and Schouten (1986). Once again, the mismatch might reflect relative displacement of western Europe with respect to the rest of Eurasia (Le Pichon et al. 1988).

9.1.4 Late to post Miocene (7-4 Ma) NW-SE compression

Compared to the stress patterns associated with the previous tectonic phases, a more complex stress configuration emerged in Late Miocene: a first-order (i.e. plate-scale) (N)NW-(S)SE compression altered by a second-order fan-shaped stress pattern around the Alps (Fig. 9.1B). The fan-shaped stress pattern might have been promoted by either indenter tectonics (e.g. Tapponnier and Molnar, 1975) or by gravitational potential forces resulting from the load of the mountain chain (e.g. Pascal and Cloetingh, 2009).

In general, Late Miocene deformation is well expressed in subalpine chains (e.g. Jura Mountains) and remains merely subtle away from the Alpine–subalpine system. Accordingly, the age of the NW-SE compression phase was chiefly determined in the Jura Mountains. The Late Miocene paleostress field reconstructed by Bergerat (1987) resembles the one recorded present-day in western-central Europe (Heidbach et al., 2018), suggesting that the current tectonic situation represents the natural prolongation of the Late Miocene phase.

FIG. 9.1, CONT'D Two main phases of the evolution of the Alpine–Mediterranean system according to Bergerat (1987) and Le Pichon et al. (1988). (A) Late Eocene (40 Ma): the orientations of maximum principal stresses are remarkably constant and in excellent agreement with the motion vectors and rotation pole of Africa with respect to Eurasia between 54 and 35 Ma as calculated by both Klitgord and Schouten (1986) and Savostin et al. (1986). (B) Late and post Miocene (7–4 Ma): maximum principal stresses are mainly oriented NW-SE in fair agreement with the rotation poles and motion vectors between 10 and 0 Ma as calculated by both Klitgord and Schouten (1986) and Savostin et al. (1986). Note the fan-shaped stress pattern around the Alps. Paleogeography and first-order structure are from Dercourt et al. (1986). AF and AR indicate rotation poles of Africa and Arabia with respect to Eurasia. Present-day coastlines are depicted as dashed lines for reference. *Redrawn from Le Pichon et al. (1988) with permission from the Geological Society of America.*

The motion vectors computed by both Klitgord and Schouten (1986) and Savostin et al. (1986) indicated displacement of Africa to the (N)NW, with respect to fixed Eurasia, in good agreement with the NW-SE compression evidenced by the study of Bergerat (1987).

Summarising, the studies of Bergerat (1987) and Le Pichon et al. (1988) showed that:

(a) paleostress determination based on fault slip data is a powerful tool to unveil the tectonic evolution of large areas and

(b) reconstructed paleostress patterns (in significantly large areas) are consistent with plate motions deduced from independent analyses.

9.2 Stress determinations from inversion of focal mechanisms

Determination of present-day stresses by means of inversion of focal mechanisms of earthquakes (see Section 4.3) is a technique that has been widely applied since the publication of the seminal paper of Angelier and Mechler (1977). The study of the East African Rift System by Delvaux and Barth (2010) is selected as an illustrative example, and the results discussed and compared with the recent geodetic measurements in Saria et al. (2014).

The East African Rift System (Fig. 9.2) is a typical example of active continental rift. It extends ~3000 km across east Africa from the Afar triangle in the north, at the triple junction with the Red Sea and the Gulf of Aden, through the highly magmatic Ethiopian Rift, the Western and Eastern rifts, which are located at the edges of the Tanzanian craton, and the Malawi Rift in the south, which is prolonged by a more diffuse deformation zone until the Mozambique coast. Offshore, the rift system appears to continue through the Mozambique Ridge, in the SW Indian Ocean, and presumably forms a triple junction with the Southwest Indian Ridge at the Andrew Bain Fault Zone (Saria et al., 2014). This incipient plate boundary separates the Nubian Plate, to the west, from the Somalian Plate, to the east (Fig. 9.2). Recent geodetic measurements are, however, in favour of a more complex plate configuration, where three minor subplates, namely Victoria, Rovuma and Lwandle subplates, are trapped between these two major plates (Saria et al., 2014).

The East African Rift System started to develop between 25 and 20 Ma with a diachronous sequence of magmatic eruptions (Ebinger, 2005). The pronounced regional domings (Nyblade and Robinson, 1994), encompassing the major segments of the rift, and the presence of high amounts of magmatic rocks, at the surface, and of slow shear wave anomalies in the mantle (Gurnis et al., 2000), at depth, suggest that rifting was triggered by active upwelling of hot mantle.

Until recently, the stresses and kinematics of the rift system have been inferred from the orientations of the major structures (Chorowicz, 2005) or from sparse fault slip measurements with rather poor timing control (Tiercelin et al., 1988). Delvaux and Barth (2010) studied 347 focal mechanisms of earthquakes recorded mainly in the East African Rift System and in other regions of Africa, occasionally (Fig. 9.2). Although all the results of their study are depicted in Fig. 9.2, the present discussion deals exclusively with those

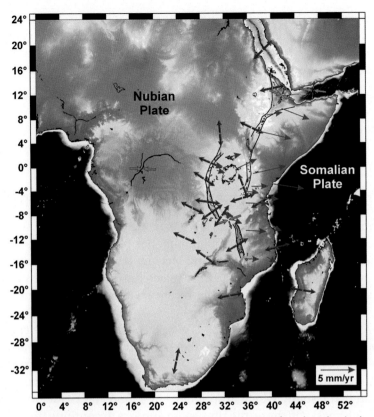

FIG. 9.2 Stress orientations determined from inversion of focal mechanisms of earthquakes in the East African Rift System and other regions of Africa (after Delvaux and Barth, 2010) compared with GPS data evidencing the displacement of the Somalian Plate relative to the Nubian Plate (after Saria et al., 2014). The main border faults of the onshore East African Rift System are schematically depicted. Divergent and convergent pairs of arrows indicate minimum and maximum principal axes of stress, respectively. *Red, green* and *blue arrows* correspond, respectively, to normal, strike-slip and reverse stress regimes. *Grey arrows* are GPS motion vectors. (For interpretation of the references to color in this figure legend, the reader is referred to the web version of this article.)

related to the rifting. For the sake of consistency, the authors divided the dataset according to rectangular geographical zones. The procedure allowed for dispatching 332 focal mechanisms within 24 geographical zones of consistent stress state. Stress tensors were computed with the inversion methods of Delvaux and Sperner (2003) and Michael (1987), for comparison. The results from the two methods were in good agreement. Finally, the quality ranking scheme of the World Stress Map project (Heidbach et al., 2018) was adapted to qualify the results.

The inversions resulted, in general, in normal stress regime with an average E-W orientation for the minimum principal stress axes (Fig. 9.2). In more detail, the minimum principal stresses were found to rotate gradually counterclockwise from ~WNW-ESE to ~WSW-ENE southwards along the rift system. Stress orientations in the Afar depression did not follow this trend but appeared intermediate between that deduced for the Ethiopian Rift, in the south, and those reconstructed for the Red Sea and the Gulf of Aden, in the north. Strike-slip stress regimes were locally determined: in the Afar in prolongation of the Aden Ridge and along the Tanganyika-Rukwa Transfer Zone, between the Western and Malawi rifts.

Delvaux and Barth (2010) concluded that the inversion of the focal mechanisms of earthquakes from the East African Rift System revealed second- and third-order stress patterns (note that first-order stress patterns stem from large-scale plate tectonics, by definition, Heidbach et al., 2018). According to the authors, the second order is generated by the gravitational potential related to the broad uplifts in eastern Africa and promotes rifting with mainly E-W extension. The third order is connected to local disturbances due to structural inheritance and is outlined by significant deviations in stress orientations with respect to the second order. To that respect, Delvaux and Barth (2010) also remarked that minimum principal axes of stress were, in general, orthogonal to major structures for most rift branches and not oblique, as previously suggested by authors assuming constant E-W orientation.

Saria et al. (2014) constructed present-day kinematic models based, in particular, on GPS velocity measurements (Fig. 9.2). Interestingly, they found that the average direction of displacement of the Somalian Plate with respect to the Nubian Plate was E-W in very good agreement with the average orientation for the minimum principal stress determined by Delvaux and Barth (2010). Fine comparison between the two sets of results is not straightforward at least because the locations they were obtained rarely coincided. Nevertheless, Saria et al. (2014) pointed out that their own prediction of almost constant E-W extension about the central part of the rift system is consistent with the detection of minor strike-slip along some faults by Delvaux and Barth (2010).

9.3 Stress magnitudes in forelands of orogens from inversion of calcite twins

This last example is taken from the review paper of Lacombe (2010). Global compilations of present-day stress measurements and indicators (Heidbach et al., 2018) reveal that average orientations of principal axes of stress inside plates remain, in general, consistent with the regimes and the orientations of stresses acting at their boundaries. The observation supports the idea that plates are sufficiently rigid to transmit stresses over long distances, a fundamental assumption in plate tectonics theory. To note, the assumption finds additional support in paleostress studies conducted at regional scales (Bergerat, 1987, see Section 9.1 and Fig. 9.1 in particular). The way stress magnitudes evolve away from plate boundaries and, therefore, the precise mechanisms assisting stress propagation within tectonic plates are, however, poorly constrained to date.

Calcite twinning paleopiezometry (see Chapter 6) provides some interesting clues on the problem. The technique was applied to map paleostresses from orogenic fronts to plate interiors in ancient forelands of North America, that is Appalachian–Ouchita (Craddock et al., 1993), Appalachian and Sevier forelands (van der Pluijm et al., 1997), and of western Europe, that is Pyrenean–Alpine foreland (Lacombe et al., 1996; Rocher et al., 2004). The main results of these studies suggest gradual decrease in differential stresses towards plate interiors (Fig. 9.3). The latter outcome appears geologically sound

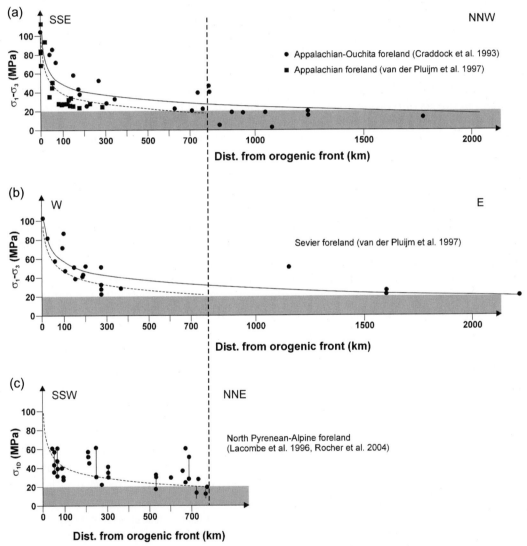

FIG. 9.3 Variation of stress magnitudes in function of distance from orogenic front as deduced from inversion of calcite twins from three forelands. (A) and (B) show decrease in differential stresses in function of distance in the Appalachian/Appalachian–Ouchita and Sevier forelands. (C) depicts decrease of the deviatoric component of σ_1, σ_{1D}, in the North Pyrenean–Alpine foreland. The *solid curves* are the exponential functions used by the authors to interpolate their respective results. The *dashed curves* represent interpolations assuming that stresses fall below a critical level (i.e. 20 MPa) at ~700 km. The *vertical dashed line* depicts the distance at which orogenic stresses fade away totally. *Redrawn from Lacombe (2010).*

and in agreement with the commonly observed decrease in both dimensions and frequency of brittle structures away from orogens.

In terms of lithosphere rheology, the results seem to confirm gradual damping of orogenic stresses by non-recoverable deformation (i.e. either brittle or ductile deformation), permitting anyway stress propagation to distances of ~1000 km inside plates. However, one has to keep in mind that the application of calcite twinning paleopiezometry is limited to the upper ~5 km of the crust and, consequently, extrapolation of its results to the whole lithosphere is not straightforward.

Considering in more detail (Fig. 9.3), the differential stress values of up to 80–100 MPa advanced by Craddock et al. (1993) and van der Pluijm (1997), for regions adjacent to orogenic fronts, appear unreasonably high for rocks at or close to the surface, even if the authors report uncertainties of ±20%. Clearly, these latter values point to exhumation from 3 to 4 km depth of the corresponding terrains encountered at the surface present-day (Lacombe, 2010). Similarly, differential stresses in the range between 20 and 30 MPa suggest ~1 km exhumation for the corresponding rocks. In brief, in the absence of further information about the paleodepths of the terrains sampled by Craddock et al. (1993) and van der Pluijm et al. (1997), it is difficult to judge on the accuracy of the stress curves they proposed. However, the fact that decrease of differential stresses is observed in the forelands, where exhumation was moderate and presumably everywhere constant, supports the validity of the decaying trend for differential stresses away from the orogenic fronts. To note, the differential stress values of less than 20 MPa reported by Craddock et al. (1993) are surprisingly low to be capable to activate any e-plane (see Chapter 6 for details).

In order to overcome the problem related to contrasting paleodepths, Lacombe et al. (1996) focused on the deviatoric component of σ_1, σ_{1D} (Fig. 9.3C). Their study was later complemented by Rocher et al. (2004). The results confirm the decreasing trend originally evidenced by Craddock et al. (1993). In particular, stresses appear to reach stable background levels of 20–30 MPa at distances of ~700 km (Fig. 9.3). We may remark that the exact form of the decaying stress function is, to date, not known and that linear interpolations may as well fit the data depicted in Fig. 9.3.

In summary, calcite twinning paleopiezometry shows stress decrease away from orogens, a result that was anticipated. However, such studies are merely rare and many aspects of the phenomenon remain unknown. As a very last remark, it is worth noting that the study on the foreland of the western Alps by Rocher et al. (2005) suggests a more complex stress curve with local ups and downs while moving to the interiors of the plate.

Further reading

Bergerat, F., 1987. Stress fields in the European platform at the time of Africa-Eurasia collision. Tectonics 6, 99–132.

Delvaux, D., Barth, A., 2010. African stress pattern from formal inversion of focal mechanism data. Tectonophysics 482, 105–128.

Lacombe, O., 2010. Calcite twins, a tool for tectonic studies in thrust belts and stable orogenic forelands. Oil Gas Sci. Technol. – Revue d'IFP Energies Nouvelles 65, 809–838.

Le Pichon, X., Bergerat, F., Roulet, M.-J., 1988. Plate kinematics and tectonics leading to the Alpine belt formation; a new analysis. Geological Society of America Special Papers. Geological Society of America, 111–132.

Saria, E., Calais, E., Stamps, D., Delvaux, D., Hartnady, C., 2014. Present-day kinematics of the East African Rift. J. Geophys. Res. Solid Earth 119, 3584–3600.

10

A practical guide to paleostress analysis

10.1 Data acquisition

It appeared to the author that this book would be incomplete without a final chapter detailing practical recommendations and guidelines to conduct a paleostress study efficiently. This chapter details partly the so-called 'tacit knowledge' normally gained while studying or working directly with experts, and is mainly aimed at students and researchers eager to grasp additional skills. Experienced colleagues will certainly find the chapter lengthy and superfluous. The recommendations listed below are inspired from the routines traditionally adopted by the 'paleostress community'. Nevertheless, the recommendations should be taken as suggestions, and the reader is most welcome to improve the procedures and to bring in her (or his) own style and personal touch of creativity.

First of all, one has to keep in mind that paleostress reconstruction methods are essentially tools for geologists and that some background in geology and moreover experience in the field are inescapable prerequisites to their use. A study based on inversion of fault slip data, gathered on the Shetland Isles, UK, is given for the sake of illustration in the following. The workflow can be applied easily to other kids of structural data (e.g. stylolithes and tension fractures) but only partly to analysis of calcite twins, which requires oriented sampling, preparation of oriented thin sections and microscope work in addition.

The tools needed for data acquisition are the traditional ones of the field geologist: hammer, lens, field book, pencils, camera, GPS, compass and, of course, a good dose of critical thinking. It is recommended to use a high-resolution geologist compass, with 1° precision, involving a built-in, high-resolution clinometer. The clinometer is particularly useful to measure pitch smoothly and accurately. It is strongly discouraged to use mobile phones for structural measurements. Mobile phones are not robust field instruments and can easily break down, and the need for charging batteries limits their use in remote areas. Additionally, as per the experience of the author, some of the available apps furnish inaccurate results (up to 20° error). Instead, it is always good to bring a spare compass, preferentially a light one, in case of problems with the main instrument.

As much as possible, one should target large outcrops (e.g. quarries, 10s to 100s metres long road cuts and cliffs), where numerous measurements can be taken in a short time, that is 100s of measurements in one day of work, typically. However, if the outcrop is relatively extended (e.g. kilometres long cliffs), it should be divided into smaller sectors with typical dimensions not exceeding ~500 m, in order to respect the conditions for stress

Paleostress Inversion Techniques. https://doi.org/10.1016/B978-0-12-811910-5.00008-7

homogeneity (see Chapter 4). Ideally, the outcrops should present differently oriented rock faces in order to avoid oversampling of the most apparent planes along one single face. If the requirement is not met, careful inspection of the outcrop helps in correction of this potential bias. Remember that paleostress reconstruction methods applied to fault slip data are theoretically valid for faults with modest (i.e. mm to cm) offsets. It is therefore recommended to treat, with much caution, measurements of faults with significant dimensions and offsets. It is also advised to avoid measuring in damage zones and cores of major faults, where large strains are clearly incompatible with the conditions of application of fault slip inversion methods, and to remain aware of potential stress deflections when measuring in the vicinities of such major faults. In brief, focus on minor fault planes that fortunately are the most common in the field and, as much as field conditions allow, stay away from complex deformation zones!

One datum consists of the attitude of the fault plane, the orientation of the stria and the sense of faulting. Depending on the type of compass and, furthermore, the habits of the geologist, strike and dip (supplemented with approximate dip direction) or dip direction and dip can be measured to quantify the attitudes of fault planes (see Section 2.5 for writing conventions). Note that for shallow dipping fault planes it is more accurate to measure striae azimuths instead of pitches.

To be more specific, a list of recommended actions to be taken in the field and particular notes and measurements that should appear in the field book is compiled below. A sample of the author's field book is also given in Fig. 10.1 as an example.

- Start with a general inspection of the outcrop to examine the main characteristics of the geology, which is exposed there, and to decide whether it is worth investing time at the place. The inspection also helps to organise the work and to detect the structures to be measured and the most accessible and safer locations to work.
- Note the coordinates of the outcrop and name the location according to local geography; names are normally easier to memorise than lists of numbers (and give a pleasant flavour of exotism...). Reporting date and other details like weather conditions help memory as well.
- A quick sketch and photos of the outcrop are particularly useful to remember both the geological context and the field conditions. More sketches should be drawn later (and photos taken) as proofs of the measured and observed structures.
- Report the encountered lithologies and formation ages; if ages are not known, it is recommended to set up a proper stratigraphic column. Stratigraphical documentation is essential to constrain potentially the respective timings of the different deformation phases when working in sedimentary environments (see an illustrative example in Kleinspehn et al., 1989).
- As already pointed out above, prefer measurement of faults with minor offsets and report approximate offsets when these are visible.
- We have seen that a collection of fault planes with diverse orientations constrains much better the inversion problem than a set of fault planes presenting monotonous attitudes (see Chapter 4). Try therefore to measure faults with markedly different

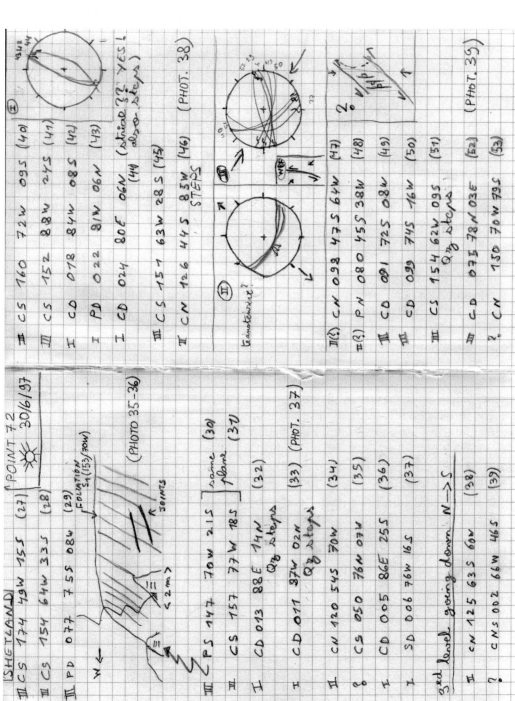

FIG. 10.1 Sample of field book presenting some of the data (mostly fault slip data) collected by the author for paleostress analyses in the Shetland Isles, UK. The data were measured in Dalradian schists exposed in the quarry of Scalloway (indicated in previous pages of the field book, see also Table 10.1 for precise coordinates of the site). The conventions used to code the fault slip data are detailed in Section 2.5 and in the text. Note the hand-made draft stereograms drawn for data consistency checking and preliminary separation (assumed deformation phases are indicated with Roman numbers) and some speculative interpretations.

strikes and dips. It might require careful inspection of the outcrop in order to detect planes that are not apparent at first glance.

- Report the types of observed kinematic indicators (see Section 2.3) and in particular, describe the minerals present on the fault planes, if any. Other qualitative descriptions of the fault surfaces (e.g. roughness, colour) might also help later during the sorting of the data.

- Carefully note absolute and relative chronology criteria (see Section 2.4). The occurrence of *syn*-sedimentary deformation is particularly important to date the reconstructed paleostress tensors (see a good example in Hippolyte et al., 1993). Field sketches and photos of the criteria are essential here to support the final interpretation of the data.

- When dealing with sediments, it is of prime importance to measure beddings where fault slip data are collected, especially if the rocks are folded or tilted. It is not uncommon that the back tilting of the faults together with the sediments hosting them brings back one of the principal axis of stress to vertical, suggesting in turn that faulting took place before folding or tilting (e.g. Fig. 4.3).

- Pay attention to and measure as well other kinds of structures (e.g. folds, veins, stylolithes…). They might furnish independent confirmation to the results of the paleostress analyses.

- Mark measurements with systematic and consistent conventions. Angelier's conventions are for example used in Fig. 10.1 (see also Section 2.5 for clarifications).

- One would like to estimate the quality of the data and its impact on the results of the inversion. Try to estimate errors, especially when you feel they might be higher than usual (e.g. measurements of poorly exposed planes), and rank the quality of the measurements. Fault slip sense can be ambiguous sometimes, it is thus recommended to set up some ranking scheme to report on the degree of uncertainty (e.g. C, P, S or U for certain, probable, supposed or unknown fault slip sense, respectively; see Fig. 10.1).

- It is particularly useful to draft preliminary stereograms and to plot the data while working in the field. Admittedly, hand-made stereograms furnish approximate representations of the data, but however present serious advantages: (1) preliminary separation of heterogeneous data, (2) immediate check of the internal consistency of the data, (3) guidance in the collection of further data (e.g. "Do I have sufficient data? Do I have enough variety in fault plane orientations?").

- As a rule of thumb, a paleostress tensor is in general well constrained with 10 to 50 fault slips, if the fault planes show enough variety. To remain on the safer side, one should aim at measuring 50 fault slip data per deformation phase and site, that is few hundreds of measurements per site are sufficient to carry out a robust study (unless the tectonic evolution of the studied area is unusually complex and involves numerous phases of deformation).

In detail, interpretation of structures and of kinematic indicators in actual field conditions, where for example exposure may be poor or rocks deeply weathered, can be arduous. The beginner may easily fall in common "field traps" if not helped by an experienced person. The reader is encouraged to carefully check the very useful accounts of Sperner and Zweigel (2010), Hippolyte et al. (2012) and Simón (2019), who warn on the danger of potential misinterpretations in the field and misuses of paleostress inversion methods in general.

To close this list of recommendations, it is advised to review the data gathered during the day when back to the base camp in the evening. Potential mistakes can be corrected easily while the memories of the day are still fresh. In addition, the reviewing of the data already at hand allows for efficient planning of the remaining field days. If time remains, one may start to tabulate and process the data with the help of a laptop and relevant computer programs.

10.2 Processing of the data

Final processing of the data is traditionally performed in the office shortly after the field campaign. Numerous fault slip inversion programs are available free of charge on internet. To the best knowledge of the author all of them deliver reliable results, it is nevertheless more comfortable to use programs that facilitate data visualisation and straightforward identification of each datum (for example the Win-TENSOR program originally introduced by Delvaux and Sperner, 2003). Also prefer mathematical methods to graphical ones, the former being more accurate (see Chapter 4).

To note, the processing of the data has to be conducted with field book at hand. Qualitative field data are the main guides of the analysis and represent primary hard data to support the final interpretation. Paleostress tensors have to be calculated site by site and, ideally, data from different sites should not be grouped.

In the (common) case of heterogeneous data, the first step of the processing consists of data separation. One may employ automatic separation of the fault slip data, if the option is available in the processing program, but the reader must be much aware that the results need to be checked meticulously (see Section 4.2.6). To the opinion of the author, the use of automatic separation is hardly justified by the relatively low amount of data, typically gathered at one site, especially when the data were already sorted out during the course of the fieldwork (see recommendations for fieldwork in Section 10.1). In addition, the thorough checking of the results after automatic separation, which involves precise identification of data associated with critical qualitative information (e.g. presence of specific minerals on the fault planes), can turn into a tedious task, nullifying the expected benefits of the automatic procedure. Manual separation, being controlled by mechanical consistency of the fault slips and qualitative observations, is strongly recommended unless the user deals with an unusually high number of data.

Low quality data might impact significantly the results of the inversion in some cases. To overcome this potential problem, most programs propose ponderation schemes, where relatively poor data (e.g. fault slips whose actual senses were difficult to determine in the field) are given less weight than good quality data in the calculations. Data ponderation presents, however, the major drawback of attributing arbitrary weights to each datum. The numerical weight of the datum can indeed overestimate or underestimate its actual quality and, ultimately, bias the results of the computations. When a mixture of data with variable quality is analysed, it is advised to conduct paleostress calculations with and without ponderation, and to compare the results to estimate the potential influence of the weighting.

Completion of the computations of the reduced paleostress tensors is relatively straightforward and guided by the numerical values returned by the minimisation function of the adopted inversion method (see Section 4.2.2). The aim of the user is to obtain paleostress determinations with reasonably low values for the minimisation function by removing successively fault slip data that appear to downgrade the results. These data are tentatively grouped in other subsets, and paleostress computations are then performed for these subsets. Data that do not fit into any of the computed tensors are finally discarded. To note, these problematic data, reflecting most probably local deformation or mistakes, represent normally a minor fraction of the whole dataset. One should raise her/his eyebrows if a large proportion of fault slip data, for example more than 10%, has been discarded.

Once the computation of all the paleostress tensors have been completed and consistent tectonic phases identified, the chronological order of events needs still to be found. Absolute timings can be derived from field evidence of *syn*-sedimentary deformation or from dating of minerals. In most cases, one must however rely on relative chronologies (e.g. cross-cutting relationships, overlapping striae or obvious re-activation of pre-existing faults; Fig. 10.2) and try to infer absolute ages from regional geology knowledge. When dealing with a significant number of tectonic phases, it is advised to adopt a systematic and objective approach to sort out the chronological order (see detailed discussion in Angelier, 1994, pp. 88–90).

The *chronology matrix method* introduced by Angelier (1991, 1994) allows for efficient classification of the tectonic phases. A square matrix containing the number of observed chronology criteria between each pair of stress events, where lines and columns correspond to 'younger' and 'older' events, respectively, is built (Fig. 10.3). As such, the diagonal of the matrix remains automatically empty (i.e. there is no chronology between one event and itself). Finding the most probable chronological order of all events consists in filling all cases of the upper-right half of the matrix and leaving its lower-left half completely empty, in the ideal case. This is done through successive permutations of the lines of the matrix (together with their corresponding columns), which is equivalent to the classical mathematical operation of matrix triangularisation. Matrix triangularisation can be made manually, but the task might become particularly laborious for large datasets. Instead, one might prefer to write or to re-use a simple numerical routine to solve the problem. In natural cases, the lower-left triangle of the matrix might still contain few numbers after having reordered the matrix. These numbers are deemed to represent natural noise or interpretation mistakes; their occurrence in the lower-left triangle should nevertheless remain incidental to ensure the validity of the proposed chronological order.

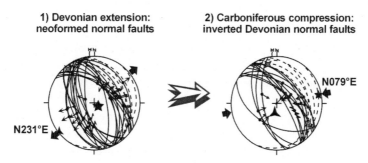

1) Devonian extension:
neoformed normal faults

2) Carboniferous compression:
inverted Devonian normal faults

N079°E

N231°E

Point 74: Brindister

FIG. 10.2 Field example of relative chronology criteria. The (quasi) Andersonian geometry of the normal faults suggests that they formed during (Devonian) extension. The reverse faults show attitudes similar to those of the normal ones and, in addition, are merely too steep for being interpreted as neoformed. The fault slip data points therefore to later inversion of the Devonian normal faults and this constrains the relative chronological order of the two events. Note well that absolute ages were inferred from the regional geological context. Planes represented in dashed lines are measured bedding attitudes, obviously the tilting of the sediments occurred prior to faulting. The collected fault planes and their respective slip vectors are projected on the lower hemisphere of a Schmidt stereogram. *Red, green* and *blue* symbols represent maximum, intermediate and minimum principal stress axes, their relative sizes suggest Φ values. *M* is magnetic north and measurements were corrected for magnetic declination prior to processing.

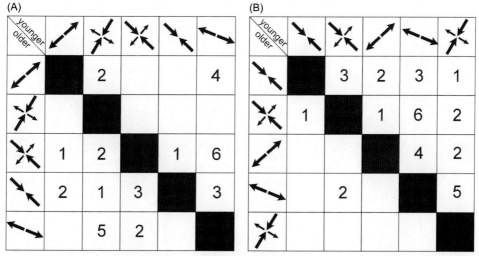

FIG. 10.3 Analysis of relative chronologies between five synthetic stress events using the chronology matrix method (Angelier, 1991, 1994). Each tectonic event is represented by combinations of arrows suggesting stress regime and orientation. The number of relative chronology criteria observed in the field between each pair of events is indicated. (A) If the chronological order of the successive tectonic events is ill-sorted, the lower-left half of the matrix is substantially filled with numbers. (B) Finding the most realistic chronological order is equivalent to reordering the matrix such as only the upper-right half of the matrix is filled in the ideal case. Residual numbers in the lower-left half should therefore represent spurious data (i.e. local effects or user's misinterpretations).

10.3 Reporting

As for other fields of science, reporting of the study results has to be clear and detailed to give the opportunity to the reader (1) to make her/his own opinion on the quality of the data and the pertinence of the interpretations and (2) to re-use the reported material in further research. The format, which is proposed here, follows this philosophy as much as possible but, admittedly, it is not unique and reflects the author's own style. The reader will find other examples for reporting in the reading list found at the end of the chapter.

We should first recall that paleostress reconstruction is above all a tool for deciphering tectonics. Representing the results on a series of maps, i.e. one map per identified tectonic phase, is therefore a natural step that any geoscientist would take. The maps should depict the determined stress regimes and orientations, using adapted and informative symbols, in the context of the regional geology; any relevant information gathered in the literature should be included (e.g. kinematics of major regional faults, Fig. 10.4).

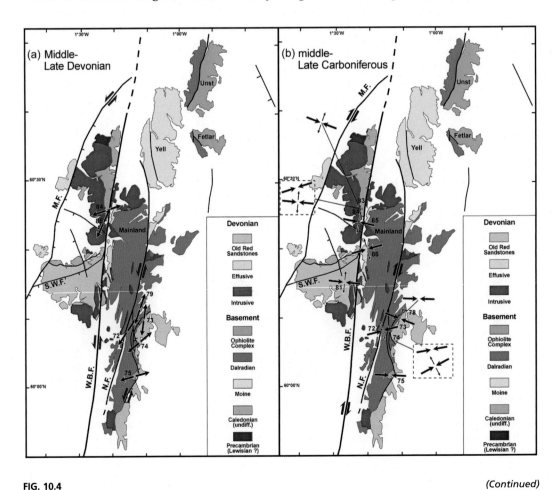

FIG. 10.4   (Continued)

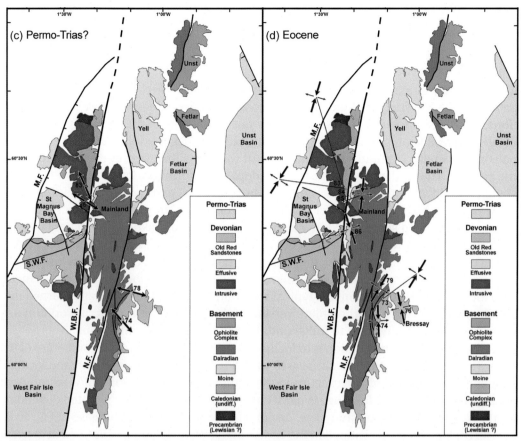

FIG. 10.4, CONT'D   Results of the paleostress analyses in the Shetland Isles, UK (author's unpublished material). The timing of each tectonic phase was inferred from previous studies in the British Isles and the North Sea. Note that the proposed Permo-Trias event remains uncertain. Diverging and converging arrows indicate normal and reverse stress regimes and combinations of both strike-slip stress regimes. Palaeozoic kinematics of the major faults are from Storetvedt (1987) and Séranne (1992). Pronounced stress rotations in Devonian are assumed to result from coeval slip along these major faults. Geological units and main structures are from the British Geological Survey Solid Edition geological map of Shetland, 1:250 000, published in 1984 (BGS, 1984). M.F., Melby Fault; N.F., Nesting Fault; S.W.F., Sulma Water Fault; W.B.F., Walls Boundary Fault. Numbers indicate studied sites (see Table 10.1 and Fig. 10.5 for full details).

Such representation of the data facilitates for the whole region under scope: (1) quick visualisation of average stress orientations in time and space, (2) straightforward identification of local stress deviations and (3) verification of the consistency between paleostresses and geological evolution (e.g. are extension directions derived from the paleostress analysis compatible with the orientations of the documented rift basins? See Fig. 10.4C). Another advantage, especially when studying large regions, is the possibility to check the coherence between the obtained stress directions and large-scale plate motions (e.g. Le Pichon et al., 1988, and Section 9.1). For a relatively high number of paleostress determinations, it is sometimes useful to create, in addition, stress trajectory maps by means of numerical interpolation of the results (see Lee and Angelier, 1994 for methodological aspects). This kind of maps allows for easy visualisation of stress perturbations and potential detection of the crustal heterogeneities that caused them (e.g. Homberg et al., 1997).

The making of maps is important, but one would like to have access to the fine details and results of the study and, in particular, to the numbers obtained after the computations. It is therefore recommended to add a table of results to the reporting. Table 10.1 details the results presented in Fig. 10.4. This table is suggested as a template for other paleostress studies and full information about the items included in it can be found in the caption. The reader is however free to adopt other formats, depending on taste, but should keep in mind that the description of the results has to be detailed and informative to be potentially useful for further analyses by colleagues. In particular, precise coordinates of the studied sites should be clearly indicated for the sake of transparency.

To be complete, the collected raw data should be presented as well. Series of stereograms depicting the fault slip data together with the main results of the paleostress reconstructions (Fig. 10.5) are, of course, preferred to long lists of numbers. Such figures might load too much the publication in a typical scientific journal, but can be easily turned into supplementary material of the online version of the paper, in our modern world. In any case, they are useful in order to judge on the quality of the determinations.

The very last recommendation is to include a figure to sum up the main outcomes of the study (Fig. 10.6). Paleostress results as well as indications on timing, duration and location should be straightforward to extract from the figure. Additional relevant information from similar studies or regional geology is particularly useful and should be adopted depending on the primary goals of the study (e.g. deciphering regional tectonics in case of Fig. 10.6).

The considerations developed in the course of this chapter, and in this book in general, should convince the reader that, although paleostress analysis is a quantitative approach, traditional fieldwork based on qualitative approaches remains the backbone of the method. To conclude, I give the word to Jacques Angelier, who was one of the most prominent authorities in the field and summarised very early in the history of the development of paleostress analysis the spirit of the methods and of quantification in geology in general (Angelier, 1983):

"L'étude qualitative est plus naturaliste, plus subtile et plus riche en ses raisonnements, et bien souvent plus féconde à travail égal; de ce fait, elle inclut une plus large part d'interprétation personnelle, donc de plus grands risques d'erreurs. L'étude quantitative est plus physique, plus brutale et en apparence plus stricte car les approximations sont nettement exprimées et souvent évaluées; elle est aussi laborieuse dans l'acquisition et le traitement des données. Les principaux risques d'erreurs ne se situent pas à l'intérieur de l'analyse quantitative, que sa rigueur même rend aisément vérifiable, mais dans ses conditions d'application. Elle est subordonnée à l'étude structurale qualitative, qui lui fixe un cadre d'utilisation et qu'en retour elle précise et enrichit."

Table 10.1 Detailed results of the paleostresses study of the Shetland Isles, UK (Figs 10.4 and 10.5). N indicates the number of fault slip data; RUP and α are quality estimators of the inversions (see Section 4.2.3), in particular RUP relates to the direct inversion method, INVD, of Angelier (1990) used in this study. The tentative quality ranking, where A represents best quality, is based on the number of data, RUP values and likelihood of results. Asterisks indicate that bedding (i.e. N010°E 15°E at Bressay and N059°E 53°S at Scottle Holm) has been rotated back to horizontal before paleostress computations. Other symbols and indications are self-explanatory.

Locality	UTM 30N easting	northing	σ1 trend°	σ1 plung°	σ2 trend°	σ2 plung°	σ3 trend°	σ3 plung°	Regime	ext/comp trend°	Φ	N	RUP %	α°	Qual.	Lithol.	Form. age	Meth.
Shetland																		
POINT72 (Scalloway)	597062	6668473	075	22	290	64	171	14	SS (Car)	075	0.5	32	45	14	A	Sericto-schists	Dalradian	INVD
			224	79	125	02	035	11	N (Dev)	035	0.5	19	37	9	A			INVD
POINT73 (quarry of Hoo Fields)	600398	6670667	211	21	338	58	112	23	SS (Eoc)	211	0.1	19	41	9	A	Tecto. mel./ conglo.	Dalradian/ Middle Dev.	INVD
			290	09	048	71	197	16	SS (Car)	290	0.2	15	35	12	A			INVD
			313	76	096	12	188	08	N (Dev)	188	0.2	21	35	12	A			INVD
POINT74 (Brindister)	599467	6665163	079	07	347	16	192	72	R (Car)	079	0.8	15	31	14	A	Sandst.	Middle Dev.	INVD
			128	81	321	09	231	02	N (Dev)	231	0.3	25	28	12	A			INVD
			267	78	047	09	138	07	N (P-Tr?)	138	0.5	9	44	15	A			INVD
			062	03	298	85	152	04	SS (Car)	062	0.8	12	39	15	A			INVD
			177	22	329	66	083	10	SS (Eoc)	177	0.7	9	32	15	A			INVD
POINT75 (Thief's Hole)	598840	6655737	242	76	344	03	075	14	N (Dev)	075	0.4	19	36	16	A	Meta-volcanics	Dalradian	INVD
			092	11	001	01	265	79	R (Car)	092	0.5	7	37	18	A			INVD
POINT76* (Bressay)	605973	6670453	347	23	220	55	088	25	SS (Eoc)	347	0.4	14	32	11	A	Sandst.	Middle Dev. - Upper Dev.	INVD
POINT78* (Scottle Holm)	602410	6673199	172	88	015	01	285	01	N (P-Tr?)	285	0.5	6	36	10	B	Congl.	Middle Dev.	INVD
			091	17	187	20	324	63	R (Car)	091	0.4	6	40	19	B			INVD
POINT79 (Loch of Vaster)	598666	6676532	222	73	111	06	019	16	N (Dev)	019	0.4	8	27	12	A	Calcschists	Dalradian	INVD
			029	15	192	74	298	04	SS (Eoc)	029	0.4	9	49	15	A			INVD
POINT81 (Hulma Water)	584576	6681192	276	28	131	57	014	16	SS (Car)	276	0.5	6	33	6	B	Sandst.	Middle Dev.	INVD
POINT83 (Sinna Water)	588513	6702579	110	16	252	71	016	11	SS (Car)	110	0.4	18	35	12	A	Plutons acid	Upper Dev.	INVD
			347	72	247	03	156	18	N (P-Tr?)	156	0.3	5	31	8	A			INVD
			016	16	240	68	110	14	SS (Eoc)	016	0.5	7	33	13	B			INVD
POINT84 (Stanes Moor)	589377	6701671	095	03	195	71	004	19	R (Car)	095	0.4	28	33	16	A	Green-schists	Unknown	INVD
			072	06	164	21	327	68	R (Car)	072	0.4	17	26	13	A			INVD
			212	05	312	64	119	26	SS (Eoc)	212	0.5	9	48	22	B			INVD
			001	71	160	18	252	07	N (Dev)	252	0.6	8	32	13	B			INVD
POINT85 (Mavis Grind)	588919	6697083	277	77	115	13	025	04	N (Dev)	025	0.4	9	41	18	B	Plutons intermed.	Upper Dev.	INVD
			308	71	212	02	121	19	N (P-Tr?)	121	0.4	8	36	4	B			INVD
			005	14	258	49	106	38	SS (Eoc)	005	0.3	4	43	7	C			INVD
			281	19	090	71	190	03	SS (Car)	281	0.3	7	25	11	B			INVD
POINT86 (Voe quarry)	595482	6691056	341	04	090	34	251	11	SS (Eoc)	341	0.4	34	34	14	A	Meta-limest.	Dalradian	INVD
			254	09	037	79	163	06	SS (Car)	254	0.5	31	32	11	A			INVD

a) NE-SW to NNE-SSW extension; Middle to Late Devonian

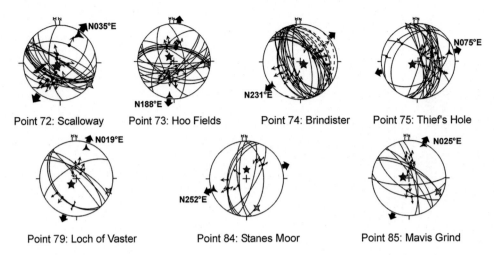

Point 72: Scalloway   Point 73: Hoo Fields   Point 74: Brindister   Point 75: Thief's Hole

Point 79: Loch of Vaster   Point 84: Stanes Moor   Point 85: Mavis Grind

b) E-W compression; middle to Late Carboniferous

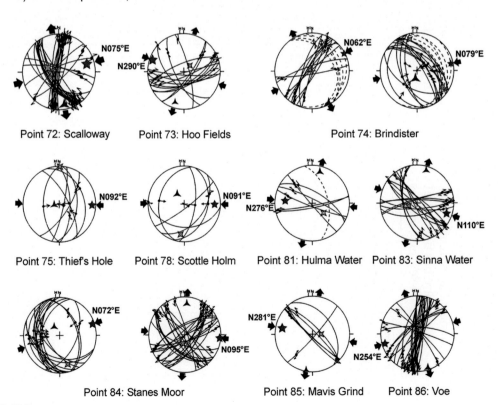

Point 72: Scalloway   Point 73: Hoo Fields   Point 74: Brindister

Point 75: Thief's Hole   Point 78: Scottle Holm   Point 81: Hulma Water   Point 83: Sinna Water

Point 84: Stanes Moor   Point 85: Mavis Grind   Point 86: Voe

FIG. 10.5

(Continued)

c) NW-SE extension; Permo-Trias?

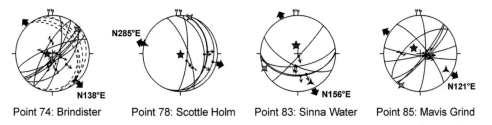

Point 74: Brindister   Point 78: Scottle Holm   Point 83: Sinna Water   Point 85: Mavis Grind

d) N-S to NNE-SSW compression; Eocene

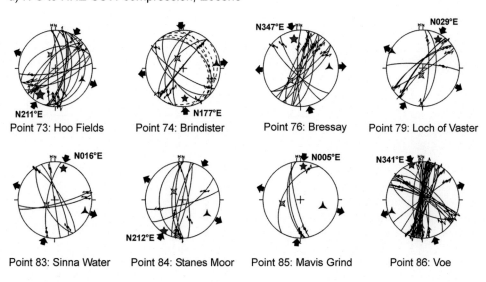

Point 73: Hoo Fields   Point 74: Brindister   Point 76: Bressay   Point 79: Loch of Vaster

Point 83: Sinna Water   Point 84: Stanes Moor   Point 85: Mavis Grind   Point 86: Voe

e) Non attributed

Point 72: Scalloway   Point 73: Hoo Fields   Point 85: Mavis Grind

FIG. 10.5, CONT'D   Paleostress results site by site and for each inferred tectonic phase (see also Fig. 10.4 and Table 10.1). The collected fault planes and their respective slip vectors are projected on the lower hemisphere of a Schmidt stereogram. *Dashed traces* indicate measured beddings. *Red, green* and *blue* symbols represent maximum, intermediate and minimum principal stress axes, their relative sizes suggest Φ values. *Arrows* depict stress regimes and compression or extension directions according to the conventions also in use in Fig. 10.4. Note in (E) the relatively small amount of data that are found incompatible with the reconstructed paleostress tensors. *M* is magnetic north and measurements were corrected for magnetic declination prior to processing.

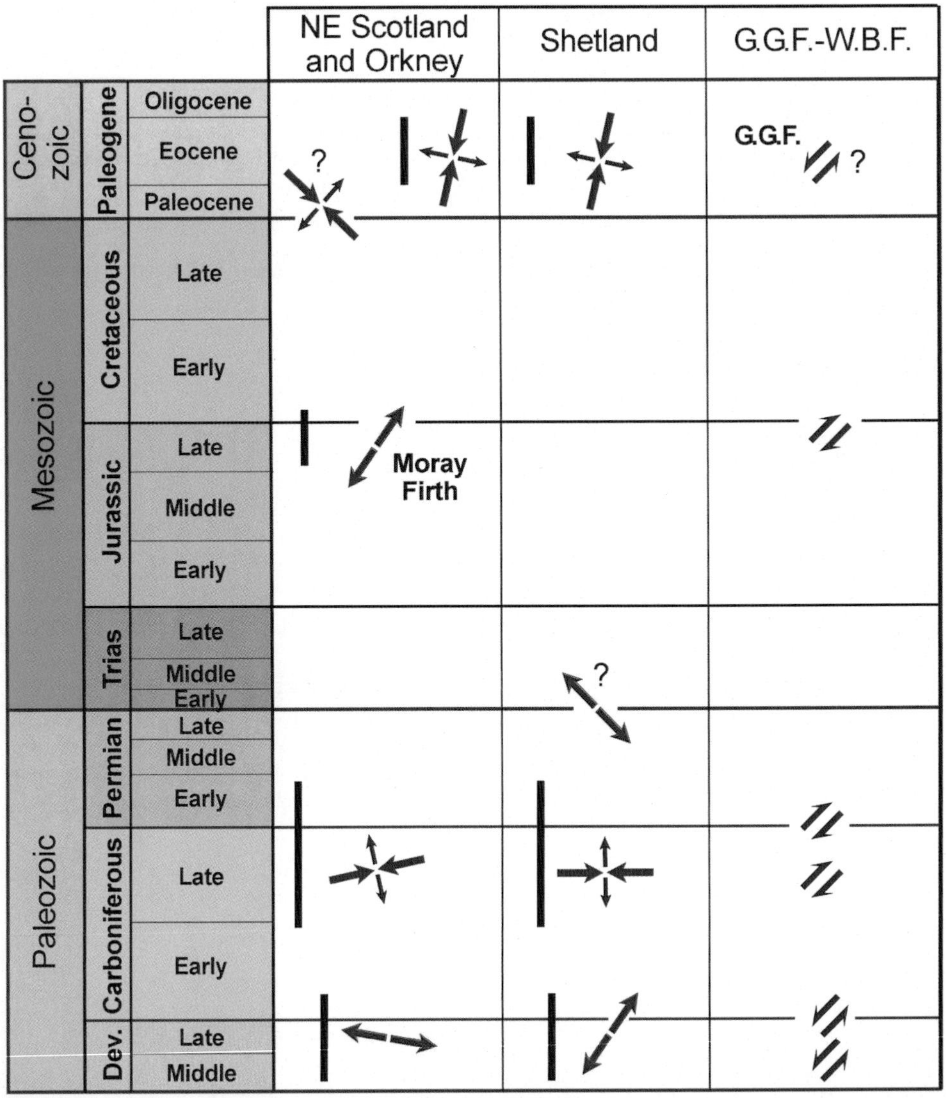

FIG. 10.6 Synthesis of the paleostress study in the Shetland Isles, UK (Figs 10.4 and 10.5, Table 10.1). *Vertical bars* indicate inferred timings and durations for each tectonic phase. *Arrows* depict stress regimes and compression or extension directions according to the conventions also in use in Fig. 10.4. The results are compared with those obtained from a similar study in NE Scotland and Orkney Isles by the author (unpublished). Periods of significant strike-slip faulting along the two major regional faults (e.g. Storetvedt, 1987; Séranne, 1992) are also reported. G.G.F., Great Glenn Fault; W.B.F., Walls Boundary Fault.

("The qualitative study is more naturalist, subtler and richer in reasonings, and very often more prolific for a similar amount of work; as such it involves a much larger part of personal interpretation and thus higher risk for mistakes. The quantitative study is more physic, more brutal and, apparently, more rigorous in the way that approximations are clearly stated and often estimated; it is also more demanding in the acquisition and processing of the data. The major risks for errors are not in the quantitative analysis, whose internal consistency allows for straightforward checks, but inside the conditions of its application. It is subordinated to the qualitative structural study, which dictates its frame of application, and in turn it [the quantitative study] enhances its precision and enriches it.")

Further reading

Angelier, J., 1994. Fault slip analysis and paleostress reconstruction. In: Hancock, P.L. (Ed.), Continental Deformation. Pergamon, Oxford, pp. 101–120.

Hippolyte, J.-C., Angelier, J., Bergerat, F., Nury, D., Guieu, G., 1993. Tectonic-stratigraphic record of paleostress time changes in the Oligocene basins of the Provence, southern France. Tectonophysics 226, 15–35.

Lamarche, J., Mansy, J.L., Bergerat, F., Averbuch, O., Hakenberg, M., Lewandowski, M., Stupnicka, E., Swidrowska, J., Wajsprych, B., Wieczorek, J., 1999. Variscan tectonics in the Holy Cross Mountains (Poland) and the role of structural inheritance during Alpine tectonics. Tectonophysics 313, 171–186.

Saintot, A., Stephens, M.B., Viola, G., Nordgulen, Ø., 2011. Brittle tectonic evolution and paleostress field reconstruction in the southwestern part of the Fennoscandian Shield, Forsmark, Sweden. Tectonics 30.

References

Aki, K., Richards, P.G., 2009. Quantitative Seismology, second ed. University Science Books, Sausalito, CA.

Amontons, G., 1699. De la résistance causée dans les machines, tant par les frottemens des parties qui les composent, que par la roideur des cordes qu'on y emploie, et la manière de calculer l'un et l'autre. Mémoires de l'Académie Royale des Sciences, Paris, 1699, in Histoire de l'Académie Royale des. Science 206–227, 1732.

Anderson, E.M., 1905. The dynamics of faulting. Trans. Edinb. Geol. Soc. 8, 387–402.

Anderson, E.M., 1938. The dynamics of sheet intrusion. Proc. Roy. Soc. Edinb. 58, 242–251.

Anderson, E.M., 1948. On lineation and petrofabric structure and the shearing movement by which they have been produced. Q. J. Geol. Soc. 104, 99–126.

Anderson, E.M., 1951. The Dynamics of Faulting and Dyke Formation with Applications to Britain. Oliver and Boyd.

André, A.-S., Sausse, J., Lespinasse, M., 2001. New approach for the quantification of paleostress magnitudes: application to the Soultz vein system (Rhine graben, France). Tectonophysics 336, 215–231.

Angelier, J., 1975. Sur l'analyse de mesures recueillies dans des sites faillés: l'utilité d'une confrontation entre les méthodes dynamique et cinématique. C. R. Acad. Sci. 281, 1805–1808.

Angelier, J., 1977. La reconstitution dynamique et géométrique de la tectonique de failles à partir de mesures locales. C. R. Acad. Sci. Sér. D 285, 637–640.

Angelier, J., 1979. Determination of the mean principal directions of stresses for a given fault population. Tectonophysics 56, T17–T26.

Angelier, J., 1983. Analyses qualitative et quantitative des populations de jeux de failles. Bull. Soc. Geol. Fr. 7 (25), 661–672.

Angelier, J., 1984. Tectonic analysis of fault slip data sets. J. Geophys. Res. Solid Earth 89, 5835–5848.

Angelier, J., 1989. From orientation to magnitudes in paleostress determinations using fault slip data. J. Struct. Geol. 11, 37–50.

Angelier, J., 1990. Inversion of field data in fault tectonics to obtain the regional stress—III. A new rapid direct inversion method by analytical means. Geophys. J. Int. 103, 363–376.

Angelier, J., 1991. Analyse chronologique matricielle et succession régionale des événements tectoniques. C. R. Acad. Sci. 312, 1633–1638.

Angelier, J., 1992a. Un élément nouveau dans la détermination des états de contrainte en tectonique cassante. Les plans de défaut non-cisaillants ou "non-failles". C. R. Acad. Sci. Sér. II 314, 381–386.

Angelier, J., 1992b. Sur l'incorporation de structures de pression et de tension dans la détermination des états de contrainte en tectonique cassante: un élargissement des méthodes d'inversion. C. R. Acad. Sci. Sér. II 314, 1233–1238.

Angelier, J., 1994. Fault slip analysis and paleostress reconstruction. In: Hancock, P.L. (Ed.), Continental Deformation. Pergamon, Oxford, pp. 53–100.

Angelier, J., 2002. Détermination du tenseur des contraintes par inversion de mécanismes au foyer de séismes sans choix de plans nodaux. Compt. Rendus Geosci. 334, 73–80.

Angelier, J., Goguel, J., 1979. Sur une méthode simple de détermination des axes principaux des contraintes pour une population de failles. C. R. Acad. Sci. Sér. D 288, 307–310.

Angelier, J., Manoussis, S., 1980. Classification automatique et distinction des phases superposées en tectonique de failles. C. R. Acad. Sci. Sér. D 290, 651–654.

Angelier, J., Mechler, P., 1977. Sur une méthode graphique de recherche des contraintes principales également utilisable en tectonique et en séismologie: la méthode des dièdres droits. Bull. Soc. Geol. Fr. 19, 1309–1318.

Angelier, J., Tarantola, A., Valette, B., Manoussis, S., 1982. Inversion of field data in fault tectonics to obtain the regional stress—I. Single phase fault populations: a new method of computing the stress tensor. Geophys. J. Int. 69, 607–621.

Armijo, R., Cisternas, A., 1978. Un problème inverse en microtectonique cassante. C. R. Acad. Sci. Sér. D 287, 595–598.

Armijo, R., Carey, E., Cisternas, A., 1982. The inverse problem in microtectonics and the separation of tectonic phases. Tectonophysics 82, 145–160.

Arthaud, F., 1969. Méthode de détermination graphique des directions de raccourcissement, d'allongement et intermédiaire d'une population de failles. Bull. Soc. Geol. Fr. 7, 729–737.

Aydin, A., Reches, Z.E., 1982. Number and orientation of fault sets in the field and in experiments. Geology 10, 107–112.

BGS, 1984. 1:250 000 Solid Edition. Shetland.

Baer, G., Beyth, M., Reches, Z., 1994. Dikes emplaced into fractured basement, Timna Igneous Complex, Israel. J. Geophys. Res. Solid Earth 99, 24039–24050.

Bahat, D., Engelder, T., 1984. Surface morphology on cross-fold joints of the Appalachian Plateau, New York and Pennsylvania. Tectonophysics 104, 299–313.

Beaudoin, N., Lacombe, O., 2018. Recent and future trends in paleopiezometry in the diagenetic domain: insights into the tectonic paleostress and burial depth history of fold-and-thrust belts and sedimentary basins. J. Struct. Geol. 114, 357–365.

Beaudoin, N., Koehn, D., Lacombe, O., Lecouty, A., Billi, A., Aharonov, E., Parlangeau, C., 2016. Fingerprinting stress: stylolite and calcite twinning paleopiezometry revealing the complexity of progressive stress patterns during folding—the case of the Monte Nero anticline in the Apennines, Italy. Tectonics 35, 1687–1712.

Bellier, O., Sebrier, M., Fourtanier, E., Gasse, F., Robles, I., 1989. Late Cenozoic evolution of the E–W striking Cajamarca deflection in the Namora Basin (Andes of Northern Peru). Ann. Tecton. 3, 77–98.

Bergerat, F., 1987. Stress fields in the European platform at the time of Africa-Eurasia collision. Tectonics 6, 99–132.

Bergerat, F., Bergues, J., Geyssant, J., 1982. Estimation des contraintes liées à la formation de décrochements dans la plateforme carbonatée de l'Allemagne du Sud. C. R. Acad. Sci. 295, 1155–1160.

Bergerat, F., Bergues, J., Geyssant, J., 1985. Estimation des contraintes liées à la formation de décrochements dans la plateforme d'Europe du Nord. Geol. Rundsch. 74, 311–320.

Bingham, C., 1974. An antipodally symmetric distribution on the sphere. Ann. Stat. 2, 1201–1225.

Bishop, A.W., 1966. The strength of soils as engineering materials. Géotechnique 16, 91–130.

Bott, M.H.P., 1959. The mechanics of oblique slip faulting. Geol. Mag. 96, 109–117.

Brillouin, L., 1946. Wave Propagation in Periodic Structures: Electric Filters and Crystal Lattices. McGraw–Hill, New York.

Burkhard, M., 1993. Calcite twins, their geometry, appearance and significance as stress-strain markers and indicators of tectonic regime: a review. J. Struct. Geol. 15, 351–368. The Geometry of Naturally Deformed Rocks.

Byerlee, J.D., 1978. Friction of rocks. Pure Appl. Geophys. 116, 615–626.

Candela, T., Renard, F., Bouchon, M., Brouste, A., Marsan, D., Schmittbuhl, J., Voisin, C., 2009. Characterization of fault roughness at various scales: implications of three-dimensional high resolution topography measurements. Pure Appl. Geophys. 166, 1817–1851.

Carey, E., 1976. Analyse Numérique d'un Modèle Mécanique Élémentaire Appliqué à l'étude d'une Population de Failles: Calcul d'un Tenseur Moyen des Contraintes à Partir des Stries de Glissement (Unpublished PhD thesis). Univ. Paris-Sud, Orsay. 138 p.

Carey, E., 1979. Recherche des directions principales de contraintes associées au jeu d'une population de failles. Rev. Géogr. Phys. Géol. Dyn. 21 (1), 57–66.

Carey, E., Brunier, B., 1974. Analyse théorique et numérique d'un modèle mécanique élémentaire appliqué à l'étude d'une population de failles. C. R. Acad. Sci. Sér. D 279, 891–894.

Carey-Gailhardis, E., Mercier, J.L., 1987. A numerical method for determining the state of stress using focal mechanisms of earthquake populations: application to Tibetan teleseisms and microseismicity of southern Peru. Earth Planet. Sci. Lett. 82, 165–179.

Célérier, B., 1988. How much does slip on a reactivated fault plane constrain the stress tensor? Tectonics 7, 1257–1278.

Célérier, B., Séranne, M., 2001. Breddin's graph for tectonic regimes. J. Struct. Geol. 23, 789–801.

Célérier, B., Etchecopar, A., Bergerat, F., Vergely, P., Arthaud, F., Laurent, P., 2012. Inferring stress from faulting: from early concepts to inverse methods. Tectonophysics 581, 206–219. Crustal Stresses, Fractures, and Fault Zones: The Legacy of Jacques Angelier.

Cheeney, R.F., 1983. Statistical Methods in Geology for Field and Lab Decisions. Allen and Unwin, London.

Choi, P.-Y., 1991. Method for determining the stress tensor using fault slip data. J. Geol. Soc. Korea 27, 383–393.

Choi, P.Y., Angelier, J., Hwang, J.H., Cadet, J.-P., Lacombe, O., Chwae, U.C., Lee, B.J., Homberg, C., Pascal, C., 1996. From shear structures to friction law: similarity of natural stress tensor in brittle tectonics. C. R. Acad. Sci. 322, 149–154.

Choi, P., Angelier, J., Cadet, J.-P., Hwang, J.-H., Sunwoo, C., 2013. Change of stress magnitudes during the polyphase tectonic history of the cretaceous Gyeongsang basin, Southeast Korea. Bull. Soc. Geol. Fr. 184, 467–484.

Chorowicz, J., 2005. The east African rift system. J. Afr. Earth Sci. 43, 79–410.

Cloos, H., 1928. Experimenten zur inneren Tektonic. Centralbl. Mineral. Pal. B 609.

Collettini, C., 2011. The mechanical paradox of low-angle normal faults: current understanding and open questions. Tectonophysics 510, 253–268.

Coulomb, C.-A., 1776. Essai sur une Application des Règles de Maximis et Minimis à Quelques Problèmes de Statique, Relatifs à L'Architecture, avec 2 Planches. Mémoires de Mathématiques et de Physique Présentés à l'Académie Royale des Sciences par Divers Savants, et lus sans ses Assemblées en 1773, vol. 7 imprimé en 1776, pp. 343–382.

Cox, S.F., Wall, V.J., Etheridge, M.A., Potter, T.F., 1991. Deformational and metamorphic processes in the formation of mesothermal vein-hosted gold deposits—examples from the Lachlan Fold Belt in Central Victoria, Australia. Ore Geol. Rev. 6, 391–423.

Craddock, J.P., Jackson, M., van Der Pluijm, B., Versical, R.T., 1993. Regional shortening fabrics in eastern North America: far-field stress transmission from the Appalachian-Ouachita orogenic belt. Tectonics 12, 257–264.

Cundall, P.A., 1971. A computer model for simulating progressive large scale movements in blocky rock systems. In: Proceedings of the Symposium of the International Society for Rock Mechanics, Society for Rock Mechanics (ISRM), France, II-8.

Curie, M., 1924. Pierre Curie. Payot, Paris.

Daubrée, A., 1879. Études Synthétiques de Géologie Expérimentale. Dunod, Paris.

Davidson, L.M., Park, R.G., 1978. Late Nagssugtoqidian stress rientation derived from deformed granodiorite dykes north of Holsteinsborg, West Greenland. J. Geol. Soc. 135, 283–289.

Delaney, P.T., Pollard, D.D., Ziony, J.I., McKee, E.H., 1986. Field relations between dikes and joints: emplacement processes and paleostress analysis. J. Geophys. Res. Solid Earth 91, 4920–4938.

Delvaux, D., Barth, A., 2010. African stress pattern from formal inversion of focal mechanism data. Tectonophysics 482, 105–128.

Delvaux, D., Sperner, B., 2003. New aspects of tectonic stress inversion with reference to the TENSOR program. Geol. Soc. Lond., Spec. Publ. 212, 75–100.

Dempster, A.P., Laird, N.M., Rubin, D.B., 1977. Maximum likelihood from incomplete data via the EM algorithm. J. R. Stat. Soc. Ser. B 39, 1–38.

DePaor, D.G., 1990. The theory of shear stress and shear strain on planes inclined to the principal directions. J. Struct. Geol. 12, 923–927.

Dercourt, J., Zonenshain, L.P., Ricou, L.-E., Kazmin, V.G., Le Pichon, X., Knipper, A.L., Grandjacquet, C., Sbortshikov, I.M., Geyssant, J., Lepvrier, C., Pechersky, D.H., Boulin, J., Sibuet, J.-C., Savostin, L.A., Sorokhtin, O., Westphal, M., Bazhenov, M.L., Lauer, J.P., Biju-Duval, B., 1986. Geological evolution of the tethys belt from the atlantic to the pamirs since the LIAS. Tectonophysics 123, 241–315. Evolution of the Tethys.

Diday, E., 1971. Une nouvelle méthode en classification automatique et reconnaissance des formes la méthode des nuées dynamiques. Rev. Stat. Appl. 19, 19–33.

Doblas, M., 1998. Slickenside kinematic indicators. Tectonophysics 295, 187–197.

Dupin, J.M., Sassi, W., Angelier, J., 1993. Homogeneous stress hypothesis and actual fault slip: a distinct element analysis. J. Struct. Geol. 15, 1033–1043.

Durelli, A.J., Phillips, E.A., Tsao, C.H., 1958. Introduction to the Theoretical and Experimental Analysis of Stress and Strain. McGraw-Hill, New York.

Ebinger, C., 2005. Continental break-up: the East African perspective. Astron. Geophys. 46, 2.16–2.21.

Ebner, M., Koehn, D., Toussaint, R., Renard, F., Schmittbuhl, J., 2009. Stress sensitivity of stylolite morphology. Earth Planet. Sci. Lett. 277, 394–398.

Ebner, M., Toussaint, R., Schmittbuhl, J., Koehn, D., Bons, P., 2010. Anisotropic scaling of tectonic stylolites: a fossilized signature of the stress field? J. Geophys. Res. Solid Earth 115.

Efron, B., 1979. Bootstrap methods: another look at the jackknife. Ann. Stat. 7, 1–26.

Engelder, T., 1999. Transitional-tensile fracture propagation: a status report. J. Struct. Geol. 21, 1049–1055.

Etchecopar, A., 1984. Étude des états de Contraintes en Tectonique Cassante et Simulation de Déformations Plastiques (Approche Mathématique) (Thèse de Doctorat-ès-Sciences). Univ. Sciences et Techniques du Languedoc, Montpellier. 270 p., unpublished.

Etchecopar, A., Vasseur, G., Daignieres, M., 1981. An inverse problem in microtectonics for the determination of stress tensors from fault striation analysis. J. Struct. Geol. 3, 51–65.

Fisher, R.A., 1921. On the "probable error" of a coefficient of correlation deduced from a small sample. Metron 1, 3–32.

Fisher, R.A., 1922. On the mathematical foundations of theoretical statistics. Phil. Trans. R. Soc. A 222, 309–368.

Fleischmann, K.H., 1992. A graphical construction for shear stress on a fault surface. J. Struct. Geol. 14, 499–502.

Fleischmann, K.H., Nemcok, M., 1991. Paleostress inversion of fault-slip data using the shear stress solution of Means (1989). Tectonophysics 196, 195–202.

Fossen, H., 2016. Structural Geology, second ed. Cambridge University Press.

Fry, N., 1992a. Direction of shear. J. Struct. Geol. 14, 253–255.

Fry, N., 1992b. Stress ratio determinations from striated faults: a spherical plot for cases of near-vertical principal stress. J. Struct. Geol. 14, 1121–1131.

Fry, N., 1999. Striated faults: visual appreciation of their constraint on possible paleostress tensors. J. Struct. Geol. 21, 7–21.

Fry, N., 2001. Stress space: striated faults, deformation twins, and their constraints on paleostress. J. Struct. Geol. 23, 1–9.

Galindo-Zaldivar, J., González-Lodeiro, F., 1988. Faulting phase differentiation by means of computer search on a grid pattern. Ann. Tecton. 2, 90–97.

Gapais, D., Cobbold, P.R., Bourgeois, O., Rouby, D., de Urreiztieta, M., 2000. Tectonic significance of fault slip data. J. Struct. Geol. 22, 881–888.

Gaul, L., Kogl, M., Wagner, M., 2003. Boundary Element Methods for Engineers and Scientists: An Introductory Course with Advanced Topics. Springer, Berlin.

Gephart, J.W., 1990a. Stress and the direction of slip on fault planes. Tectonics 9, 845–858.

Gephart, J.W., 1990b. FMSI: a Fortran program for inverting fault/slickenside and earthquake focal mechanism data to obtain the regional stress tensor. Comput. Geosci. 16 (7), 953–989.

Gephart, J.W., 1992. Fault geometries as constraints on the stress tensor. Abstract T41C-3. EOS Trans. Am. Geophys. Union 73 (14), 298.

Gephart, J.W., Forsyth, D.W., 1984. An improved method for determining the regional stress tensor using earthquake focal mechanism data: application to the San Fernando earthquake sequence. J. Geophys. Res. Solid Earth 89, 9305–9320.

Granado, P., Tavani, S., Carrera, N., Muñoz, J.A., 2018. Deformation pattern around the Conejera fault blocks (Asturian Basin, North Iberian Margin). Geol. Acta 16, 357–373.

Griffith, A.A., 1921. The phenomena of rupture and flow in solids. Philos. Trans. R. Soc. Lond. A 221, 163–198.

Griffith, A.A., 1924. The theory of rupture. In: Proc. Int. Congr. Appl. Mech, pp. 56–63.

Guimerá, J., 1983. Évolution de la déformation alpine dans le NE de la Chaîne Ibérique et dans la Chaîne Côtière Catalane. C.R. Acad. Sci. Sér. D 297, 425–430.

Guimerá, J., 1984. Palaeogene evolution of deformation in the northeastern Iberian Peninsula. Geol. Mag. 121, 413–420.

Gurnis, M., Mitrovica, J., Ritsema, J., van Heist, H.J., 2000. Constraining mantle density structure using geological evidence of surface uplift rates: the case of the African super-plume. Geochem. Geophys. Geosyst. 1, 1020.

Gushchenko, O.I., 1975. Kinematical principle of reconstructing the principal stress directions (from geological and seismic data). Dokl. Akad. Nauk SSSR 225, 557–560 (in Russian).

Gushchenko, O.I., 1979. The method of the kinematic analysis of structures of destruction at reconstruction of fields of tectonic stresses. In: Fields of Stress a Lithosphere. Nauka, Moscow, pp. 7–25 (in Russian).

Gzovskii, M.V., 1954. Tectonic stress fields. Izv. Akad. Nauk. SSSR Ser. Geogr. 3, 390–410 (in Russian).

Hancock, P.L., 1985. Brittle microtectonics: principles and practice. J. Struct. Geol. 7, 437–457.

Hansen, J.-A., 2013. Direct inversion of stress, strain or strain rate including vorticity: a linear method of homogenous fault–slip data inversion independent of adopted hypothesis. J. Struct. Geol. 51, 3–13.

Hardcastle, K.C., 1989. Possible paleostress tensor configurations derived from fault-slip data in eastern Vermont and western New Hampshire. Tectonics 8, 265–284.

Hardcastle, K.C., Hills, L.S., 1991. BRUTE3 and SELECT: QUICKBASIC 4 programs for determination of stress tensor configurations and separation of heterogeneous populations of fault-slip data. Comput. Geosci. 17, 23–43.

Healy, D., Blenkinsop, T.G., Timms, N.E., Meredith, P.G., Mitchell, T.M., Cooke, M.L., 2015. Polymodal faulting: time for a new angle on shear failure. J. Struct. Geol. 80, 57–71.

Heidbach, O., Rajabi, M., Cui, X., Fuchs, K., Müller, B., Reinecker, J., Reiter, K., Tingay, M., Wenzel, F., Xie, F., Ziegler, M., Zoback, M.-L., Zoback, M., 2018. The World Stress Map database release 2016: crustal stress pattern across scales. Tectonophysics 744, 484–498.

Hippolyte, J.-C., Angelier, J., Bergerat, F., Nury, D., Guieu, G., 1993. Tectonic-stratigraphic record of paleostress time changes in the Oligocene basins of the Provence, southern France. Tectonophysics 226, 15–35.

Hippolyte, J.C., Bergerat, F., Gordon, M., Bellier, O., Espurt, N., 2012. Keys and pitfalls in mesoscale fault analysis and paleostress reconstructions, the use of Angelier's methods. Tectonophysics 581, 144–162.

Hirayama, J., Kakimi, T., 1965. An analytical method for determination of tectonic stress field and its applications. Earth Sci. 78, 23–31 (in Japanese).

Hoek, E., Brown, E.T., 1980. Underground Excavations in Rock. Institution of Mining and Metallurgy, London.

Hoeppner, R., 1955. Tektonik im Schiefergebirge. Geol. Rundsch. 44, 26–58.

Homberg, C., Hu, J.C., Angelier, J., Bergerat, F., Lacombe, O., 1997. Characterization of stress perturbation near major fault zones: insights from 2-D distinct-element numerical modelling and field studies (Jura Mountains). J. Struct. Geol. 19, 703–718.

Huang, Q., 1988. Computer-based method to separate heterogeneous sets of fault-slip data into sub-sets. J. Struct. Geol. 10, 297–299.

Huang, Q., Angelier, J., 1989. Inversion of field data in fault tectonics to obtain the regional stress—II. Using conjugate fault sets within heterogeneous families for computing palaeostress axes. Geophys. J. Int. 96, 139–149.

Hubbert, M.K., Rubey, W.W., 1959. Role of fluid pressures in mechanics of overthrust faulting: I. mechanics of fluid-filled porous solids and its application to overthrust faulting. GSA Bull. 70, 115–166.

Inglis, C.E., 1913. Stresses in plates due to the presence of cracks and sharp corners. Trans. Inst. Naval Arch. 55, 219–241.

Jaeger, J.C., Cook, N.G.W., Zimmerman, R., 2007. Fundamentals of Rock Mechanics. Wiley.

Jamison, W.R., Spang, J.H., 1976. Use of calcite twin lamellae to infer differential stress. GSA Bull. 87, 868–872.

Jaques, L., Pascal, C., 2017. Full paleostress tensor reconstruction using quartz veins of Panasqueira Mine, central Portugal; part I: paleopressure determination. J. Struct. Geol. 102, 58–74.

Jolly, R.J.H., Sanderson, D.J., 1997. A Mohr circle construction for the opening of a pre-existing fracture. J. Struct. Geol. 19, 887–892.

Kaven, J.O., Maerten, F., Pollard, D.D., 2011. Mechanical analysis of fault slip data: implications for paleostress analysis. J. Struct. Geol. 33, 78–91.

Kleinspehn, K.L., Pershing, J., Teyssier, C., 1989. Paleostress stratigraphy: a new technique for analyzing tectonic control on sedimentary-basin subsidence. Geology 17, 253–256.

Klitgord, K.D., Schouten, H., 1986. Plate Kinematics of the Central Atlantic. In: Vogt, P.R., Tucholke, B.E. (Eds.), The geology of North America v. M, The western North Atlantic region. Geological Society of America: Decade of North American Geology, pp. 351–378.

Krantz, R.W., 1988. Multiple fault sets and three-dimensional strain: theory and application. J. Struct. Geol. 10, 225–237.

Lacombe, O., 2010. Calcite twins, a tool for tectonic studies in thrust belts and stable orogenic forelands. Oil Gas Sci. Technol. 65, 809–838.

Lacombe, O., 2012. Do fault slip data inversions actually yield "paleostresses" that can be compared with contemporary stresses? A critical discussion. Compt. Rendus Geosci. 344, 159–173.

Lacombe, O., Angelier, J., Laurent, P., Bergerat, F., Tourneret, C., 1989. Contribution de l'analyse des macles de la calcite à la connaissance de l'histoire tectonique d'une région: l'exemple de la plate-forme carbonatée bourguignonne. C. R. Acad. Sci. 309 (II), 1979–1984.

Lacombe, O., Angelier, J., Laurent, P., Bergerat, F., Tourneret, C., 1990. Joint analyses of calcite twins and fault slips as a key for deciphering polyphase tectonics: Burgundy as a case study. Tectonophysics 182, 279–300. Paleomagnetic Constraints on Crustal Motions.

Lacombe, O., Laurent, P., Rocher, M., 1996. Magnitude de la contrainte déviatorique pyrénéenne dans l'avant-pays nord pyrénéen. C.R. Acad. Sci. II 322, 229–235.

Lamarche, J., Mansy, J.L., Bergerat, F., Averbuch, O., Hakenberg, M., Lewandowski, M., Stupnicka, E., Swidrowska, J., Wajsprych, B., Wieczorek, J., 1999. Variscan tectonics in the Holy Cross Mountains (Poland) and the role of structural inheritance during Alpine tectonics. Tectonophysics 313, 171–186.

Laubach, S.E., 1989. Paleostress directions from the preferred orientation of closed microfractures (fluid-inclusion planes) in sandstone, East Texas basin, U.S.A. J. Struct. Geol. 11, 603–611.

Laurent, P., 1984. Les Macles de la Calcite en Tectonique: Nouvelles Méthodes Dynamiques et Premières Applications (Thèse de Doctorat-ès-Sciences). Univ. Sciences et Techniques du Languedoc, Montpellier. 324 p.

Laurent, P., Bernard, P., Vasseur, G., Etchecopar, A., 1981. Stress tensor determination from the study of e twins in calcite: a linear programming method. Tectonophysics 78, 651–660. The Effect of Deformation on Rocks.

Laurent, P., Tourneret, C., Laborde, O., 1990. Determining deviatoric stress tensors from calcite twins: applications to monophased synthetic and natural polycrystals. Tectonics 9, 379–389.

Le Pichon, X., Bergerat, F., Roulet, M.-J., 1988. Plate kinematics and tectonics leading to the Alpine belt formation; A new analysis. In: Geological Society of America Special Papers. Geological Society of America, pp. 111–132.

Lee, J.-C., Angelier, J., 1994. Paleostress trajectory maps based on the results of local determinations: the "lissage" program. Comput. Geosci. 20, 161–191.

Lejri, M., 2015. Subsurface Stress Inversion Modeling Using Linear Elasticity: Sensitivity Analysis and Applications (PhD thesis). Montpellier University.

Lejri, M., Maerten, F., Maerten, L., Soliva, R., 2015. Paleostress inversion: a multiparametric geomechanical evaluation of the Wallace-Bott assumptions. Tectonophysics 657, 129–143.

Lejri, M., Maerten, F., Maerten, L., Soliva, R., 2017. Accuracy evaluation of both Wallace-Bott and BEM-based paleostress inversion methods. Tectonophysics 694, 130–145.

Lepvrier, C., Martínez-García, E., 1990. Fault development and stress evolution of the post-Hercynian Asturian Basin (Asturias and Cantabria, Northwest Spain). Tectonophysics 184, 345–356.

Lepvrier, C., Mougenot, D., 1984. Déformations cassantes et champs de contrainte post-hercyniens dans l'Ouest de l'Ibérie (Portugal). Rev. Géol. Dynam. Géog. Phys. 25, 291–305.

Lespinasse, M., 1999. Are fluid inclusion planes useful in structural geology? J. Struct. Geol. 21, 1237–1243.

Lespinasse, M., Cathelineau, M., 1995. Paleostress magnitudes determination by using fault slip and fluid inclusions planes data. J. Geophys. Res. Solid Earth 100, 3895–3904.

Lespinasse, M., Pêcher, A., 1986. Microfracturing and regional stress field: a study of the preferred orientations of fluid-inclusion planes in a granite from the Massif Central, France. J. Struct. Geol. 8, 169–180.

Letouzey, J., Trémolières, P., 1980. Paleo-stress fields around the Mediterranean since the Mesozoic from microtectonics: comparison with plate tectonic data. Rock Mech. 9, 173–192.

Liesa, C.L., Lisle, R.J., 2004. Reliability of methods to separate stress tensors from heterogeneous fault-slip data. J. Struct. Geol. 26, 559–572.

Lisle, R.J., 1987. Principal stress orientations from faults: an additional constraint. Ann. Tecton. 1, 155–158.

Lisle, R.J., 1988. Romsa: a basic program for paleostress analysis using fault-striation data. Comput. Geosci. 14, 255–259.

Lisle, R.J., 1989. A simple construction for shear stress. J. Struct. Geol. 11, 493–495.

Lisle, R.J., 1992. New method of estimating regional stress orientations: application to focal mechanism data of recent British earthquakes. Geophys. J. Int. 110, 276–282.

Lisle, R.J., 1998. Simple graphical construction for the direction of shear. J. Struct. Geol. 20, 969–973.

Lisle, R.J., 2013. A critical look at the Wallace-Bott hypothesis in fault-slip analysis. Bull. Soc. Geol. Fr. 184, 299–306.

Lisle, R.J., Srivastava, D.C., 2004. Test of the frictional reactivation theory for faults and validity of fault-slip analysis. Geology 32, 569–572.

Lisle, R.J., Vandycke, S., 1996. Separation of multiple stress events by fault striation analysis: an example from Variscan and younger structures at Ogmore, South Wales. J. Geol. Soc. 153, 945–953.

Lisle, R., Orife, T., Arlegui, L., 2001. A stress inversion method requiring only fault slip sense. J. Geophys. Res. Solid Earth 106, 2281–2289.

Lisle, R.J., Orife, T.O., Arlegui, L., Liesa, C., Srivastava, D.C., 2006. Favoured states of palaeostress in the Earth's crust: evidence from fault-slip data. J. Struct. Geol. 28, 1051–1066. New Dynamics in Palaeostress Analysis.

Love, J.J., 2007. Bingham statistics. In: Gubbins, D., Herrero-Bervira, E. (Eds.), Encyclopedia of Geomagnetism and Paleomagnetism. Springer, Dordrecht, pp. 45–47.

Maerten, L., 2000. Variation in slip on intersecting normal faults: implications for paleostress inversion. J. Geophys. Res. Solid Earth 105, 25553–25565.

Maerten, F., 2010. Geomechanics to Solve Geological Structure Issues: Forward, Inverse and Restoration Modeling (PhD thesis). Montpellier University.

Maerten, F., Maerten, L., Pollard, D.D., 2014. iBem3D, a three-dimensional iterative boundary element method using angular dislocations for modeling geologic structures. Comput. Geosci. 72, 1–17.

Maerten, L., Maerten, F., Lejri, M., Gillespie, P., 2016a. Geomechanical paleostress inversion using fracture data. J. Struct. Geol. 84, 197–213.

Maerten, F., Madden, B., Pollard, D.D., Maerten, L., 2016b. Incorporating fault mechanics into inversions of aftershock data for the regional remote stress, with application to the 1992 Landers, California earthquake. Tectonophysics 674, 52–64.

Mandelbrot, B.B., 1985. Self-affine fractals and fractal dimension. Phys. Scr. 32, 257–260.

Marrett, R., Allmendinger, R.W., 1990. Kinematic analysis of fault-slip data. J. Struct. Geol. 12, 973–986.

Martínez-Garzón, P., Kwiatek, G., Bohnhoff, M., Dresen, G., 2016. Impact of fluid injection on fracture reactivation at the Geysers geothermal field. J. Geophys. Res. Solid Earth 121, 7432–7449.

Mattauer, M., Mercier, J.L., 1980. Microtectonique et grande tectonique. Livre jubilaire du Cent Cinquantenaire 1830–1980. Mem. H.S. Soc. Geol. Fr. 10, 141–161.

Mazzarini, F., Isola, I., 2007. Hydraulic connection and fluid overpressure in upper crustal rocks: evidence from the geometry and spatial distribution of veins at botrona quarry, Southern Tuscany, Italy. J. Struct. Geol. 29, 1386–1399.

McClintock, F.A., Walsh, J.B., 1962. Friction of Griffith cracks in rock under pressure. In: Proc. Fourth U.S. Congr. Appl. Mech. American Society of Mechanical Engineers, Berkeley, pp. 1015–1021.

McFarland, J.M., Morris, A.P., Ferrill, D.A., 2012. Stress inversion using slip tendency. Comput. Geosci. 41, 40–46.

McKeagney, C.J., Boulter, C.A., Jolly, R.J.H., Foster, R.P., 2004. 3-D Mohr circle analysis of vein opening, Indarama lode-gold deposit, Zimbabwe: implications for exploration. J. Struct. Geol. 26, 1275–1291.

McKenzie, D.P., 1969. The relation between fault plane solutions for earthquakes and the directions of the principal stresses. Bull. Seismol. Soc. Am. 59, 591–601.

Means, W.D., 1976. Stress and Strain: Basic Concepts of Continuum Mechanics for Geologists. Springer-Verlag, New York.

Means, W.D., 1989. A construction for shear stress on a generally-oriented plane. J. Struct. Geol. 11, 625–627.

Michael, A.J., 1984. Determination of stress from slip data: faults and folds. J. Geophys. Res. Solid Earth 89, 11517–11526.

Michael, A.J., 1987. Use of focal mechanisms to determine stress: a control study. J. Geophys. Res. 92 (B), 357–368.

Mohr, O., 1900. Welche Umstände bedingen die Elastizitätsgrenze und den Bruch eines Materials? Z. Ver. Dtsch. Ing. 24 (45), 1524–1530. und Z. Ver. Dtsch. Ing. 24, 46, 1572–1577.

Morris, A., Ferrill, D.A., Henderson, D.B., 1996. Slip-tendency analysis and fault reactivation. Geology 24, 275–278.

Morris, A.P., Ferrill, D.A., McGinnis, R.N., 2016. Using fault displacement and slip tendency to estimate stress states. J. Struct. Geol. 83, 60–72.

Mostafa, M.E., 2005. Iterative direct inversion: an exact complementary solution for inverting fault-slip data to obtain palaeostresses. Comput. Geosci. 31, 1059–1070.

Navier, C.L., 1833. Résumé des leçons données à l'École des ponts et chaussées sur l'application de la mécanique à l'Établissement des constructions et des machines, tome 1, Première partie contenant des leçons sur la résistance des matériaux, et sur l'établissement des constructions en terre, en maçonnerie et en charpente, second ed. Carilian-Gœury, Paris.

Nelder, J.A., Mead, R., 1965. A simplex method for function minimization. Comput. J. 7, 308–313.

Nemcok, M., Lisle, R.J., 1995. A stress inversion procedure for polyphase fault/slip data sets. J. Struct. Geol. 17, 1445–1453.

Nemcok, M., Kovac, D., Lisle, R.J., 1999. A stress inversion procedure for polyphase calcite twin and fault/slip data sets. J. Struct. Geol. 21, 597–611.

Nieto-Samaniego, A.F., Alaniz-Alvarez, S.A., 1997. Origin and tectonic interpretation of multiple fault patterns. Tectonophysics 270, 197–206.

Nigon, B., Englert, A., Pascal, C., Saintot, A., 2017. Multi-scale characterization of joint surface roughness. J. Geophys. Res. Solid Earth 122, 9714–9728.

Nyblade, A.A., Robinson, S.W., 1994. The African superswell. Geophys. Res. Lett. 21, 765–768.

Orife, T., Lisle, R.J., 2003. Numerical processing of palaeostress results. J. Struct. Geol. 25, 949–957.

Orife, T., Arlegui, L., Lisle, R.J., 2002. DIPSLIP: a QuickBasic stress inversion program for analysing sets of faults without slip lineations. Comput. Geosci. 28, 775–781.

Otsubo, M., Yamaji, A., 2006. Improved resolution of the multiple inverse method by eliminating erroneous solutions. Comput. Geosci. 32, 1221–1227.

Otsubo, M., Sato, K., Yamaji, A., 2006. Computerized identification of stress tensors determined from heterogeneous fault-slip data by combining the multiple inverse method and k-means clustering. J. Struct. Geol. 28, 991–997. New Dynamics in Palaeostress Analysis.

Park, W.C., Schot, E.H., 1968. Stylolites; their nature and origin. J. Sediment. Res. 38, 175–191.

Parlangeau, C., Lacombe, O., Schueller, S., Daniel, J.-M., 2018. Inversion of calcite twin data for paleostress orientations and magnitudes: a new technique tested and calibrated on numerically-generated and natural data. Tectonophysics 722, 462–485.

Pascal, C., 2002. Interaction of faults and perturbation of slip: influence of anisotropic stress states in the presence of fault friction and comparison between Wallace-Bott and 3D distinct element models. Tectonophysics 356, 307–322.

Pascal, C., 2004. SORTAN, a Unix program for calculation and graphical presentation of fault slip as induced by stresses. Comput. Geosci. 30, 259–265.

Pascal, C., Angelier, J., 2003. SORTAN an analytical method to determine fault slip as induced by stresses. Math. Geol. 35, 627–642.

Pascal, C., Cloetingh, S.A.P.L., 2009. Gravitational potential stresses and stress field of passive continental margins: insights from the South-Norway shelf. Earth Planet. Sci. Lett. 277, 464–473.

Pascal, C., Angelier, J., Seland, R., Lepvrier, C., 2002. A simplified model of stress-slip relationships: application to the Frøy Field, northern North Sea. Tectonophysics 357, 103–118.

Passchier, C.W., Trouw, R.A.J., 2005. Microtectonics, second ed. Springer Verlag, Berlin.

Petit, J.P., 1987. Criteria for the sense of movement on fault surfaces in brittle rocks. J. Struct. Geol. 9, 597–608.

Petit, J.-P., Barquins, M., 1988. Can natural faults propagate under mode II conditions? Tectonics 7, 1243–1256.

Pfiffner, A., Burkhard, M., 1987. Determination of paleostress axis orientations from fault, twin and earthquake data. Ann. Tecton. 1, 48–57.

Phan-Trong, T., 1993. An inverse problem for the determination of the stress tensor from polyphased fault sets and earthquake focal mechanisms. Tectonophysics 224, 393–411.

Pollard, D.D., Aydin, A., 1988. Progress in understanding jointing over the past century. Geol. Soc. Am. Bull. 100, 1181–1204.

Pollard, P.D., Fletcher, R.C., 2005. Fundamentals of Structural Geology. Cambridge University Press.

Pollard, D.D., Saltzer, S.D., Rubin, A.M., 1993. Stress inversion methods: are they based on faulty assumptions? J. Struct. Geol. 15, 1045–1054.

Popova, E., Popov, V.L., 2015. The research works of Coulomb and Amontons and generalized laws of friction. Friction 3, 183–190.

Ragan, D.M., 1990. Direction of shear. J. Struct. Geol. 12, 929–931.

Ramsay, J., 1980. The crack–seal mechanism of rock deformation. Nature 284, 135–139.

Ramsay, J.G., Huber, M.I., 1987. The Techniques of Modern Structural Geology: Strain Analysis. Academic Press, London.

Ramsay, J.G., Lisle, R.J., 2000. The techniques of modern structural geology. In: Applications of Continuum Mechanics in Structural Geology. vol. 3. Academic Press, London.

Ramsey, J., Chester, F., 2004. Hybrid fracture and the transition from extension fracture to shear fracture. Nature 428, 63–66.

Reches, Z., 1987. Determination of the tectonic stress tensor from slip along faults that obey the Coulomb yield condition. Tectonics 6, 849–861.

Reches, Z.E., Dieterich, J.H., 1983. Faulting of rocks in three-dimensional strain fields I. Failure of rocks in polyaxial, servo-control experiments. Tectonophysics 95, 111–132.

Reches, Z., Baer, G., Hatzor, Y., 1992. Constraints on the strength of the upper crust from stress inversion of fault slip data. J. Geophys. Res. Solid Earth 97, 12481–12493.

Renard, F., Schmittbuhl, J., Gratier, J.-P., Meakin, P., Merino, E., 2004. Three-dimensional roughness of stylolites in limestones. J. Geophys. Res. Solid Earth 109.

Riedel, W., 1929. Zur Mechanik Geologischer Brucherscheinungen. Zentbl. Miner. Geol. Paläeont. B, 354–368.

Rispoli, R., 1981. Stress fields about strike-slip faults inferred from stylolites and tension gashes. Tectonophysics 75, 29–36.

Rispoli, R., Vasseur, G., 1983. Variation with depth of the stress tensor anisotropy inferred from microfault analysis. Tectonophysics 93, 169–184.

Ritz, J.F., 1994. Determining the slip vector by graphical construction: use of a simplified representation of the stress tensor. J. Struct. Geol. 16, 737–741.

Rocher, M., Cushing, M., Lemeille, F., Lozac'h, Y., Angelier, J., 2004. Intraplate paleostresses reconstructed with calcite twinning and faulting: improved method and application to the Lorraine platform area (eastern France). Tectonophysics 387, 1–21.

Rocher, M., Cushing, M., Lemeille, F., Baize, S., 2005. Stress induced by the Mio-Pliocene Alpine collision in northern France. Bull. Soc. Geol. Fr. 176, 319–328.

Roedder, E., 1984. Fluid inclusions. In: Reviews in Mineralogy. vol. 12. Mineralogical Society of America. 644 pp.

Rolland, A., Toussaint, R., Baud, P., Schmittbuhl, J., Conil, N., Koehn, D., Renard, F., Gratier, J.-P., 2012. Modeling the growth of stylolites in sedimentary rocks. J. Geophys. Res. Solid Earth 117, B06403.

Rolland, A., Toussaint, R., Baud, P., Conil, N., Landrein, P., 2014. Morphological analysis of stylolites for paleostress estimation in limestones. Int. J. Rock Mech. Min. Sci. 67, 212–225.

Rowe, K.J., Rutter, E.H., 1990. Palaeostress estimation using calcite twinning: experimental calibration and application to nature. J. Struct. Geol. 12, 1–17.

Saintot, A., Stephens, M.B., Viola, G., Nordgulen, Ø., 2011. Brittle tectonic evolution and paleostress field reconstruction in the southwestern part of the Fennoscandian Shield, Forsmark, Sweden. Tectonics 30.

Saria, E., Calais, E., Stamps, D., Delvaux, D., Hartnady, C., 2014. Present-day kinematics of the East African Rift. J. Geophys. Res. Solid Earth 119, 3584–3600.

Sassi, W., Carey-Gailhardis, E., 1987. Interprétation mécanique du glissement sur les failles: introduction d'un critère de frottement. Ann. Tecton., 139–154.

Sasvári, Á., Baharev, A., 2014. SG2PS (structural geology to postscript converter)—a graphical solution for brittle structural data evaluation and paleostress calculation. Comput. Geosci. 66, 81–93.

Sato, K., 2006. Incorporation of incomplete fault-slip data into stress tensor inversion. Tectonophysics 421, 319–330.

Sato, K., Yamaji, A., 2006a. Uniform distribution of points on a hypersphere for improving the resolution of stress tensor inversion. J. Struct. Geol. 28, 972–979. New Dynamics in Palaeostress Analysis.

Sato, K., Yamaji, A., 2006b. Embedding stress difference in parameter space for stress tensor inversion. J. Struct. Geol. 28, 957–971. New Dynamics in Palaeostress Analysis.

Sato, K., Yamaji, A., Tonai, S., 2013. Parametric and non-parametric statistical approaches to the determination of paleostress from dilatant fractures: application to an early Miocene dike swarm in Central Japan. Tectonophysics 588, 69–81.

Savostin, L.A., Sibuet, J.-C., Zonenshain, L.P., Le Pichon, X., Roulet, M.-J., 1986. Kinematic evolution of the Tethys belt from the Atlantic Ocean to the pamirs since the Triassic. Tectonophysics 123, 1–35. Evolution of the Tethys.

Schmittbuhl, J., Schmitt, F., Scholz, C., 1995. Scaling invariance of crack surfaces. J. Geophys. Res. Solid Earth 100, 5953–5973.

Schmittbuhl, J., Renard, F., Gratier, J.P., Toussaint, R., 2004. Roughness of stylolites: implications of 3D high resolution topography measurements. Phys. Rev. Lett. 93, 238501.

Schwarz, G.E., 1978. Estimating the dimension of a model. Ann. Stat. 6, 461–464.

Secor, D., 1965. Role of fluid pressure in jointing. Am. J. Sci. 263, 633–646.

Segall, P., Pollard, D., 1980. Mechanics of discontinuous faulting. J. Geophys. Res. 85, 4337–4350.

Séranne, M., 1992. Devonian extensional tectonics versus carboniferous inversion in the northern Orcadian basin. J. Geol. Soc. Lond. 149, 27–37.

Shan, Y., Fry, N., 2005. A hierarchical cluster approach for forward separation of heterogeneous fault/slip data into subsets. J. Struct. Geol. 27, 929–936.

Shan, Y., Suen, H., Lin, G., 2003. Separation of polyphase fault/slip data: an objective-function algorithm based on hard division. J. Struct. Geol. 25, 829–840.

Shan, Y., Li, Z., Lin, G., 2004. A stress inversion procedure for automatic recognition of polyphase fault/slip data sets. J. Struct. Geol. 26, 919–925.

Shan, Y., Lin, G., Li, Z., Zhao, C., 2006. Influence of measurement errors on stress estimated from single-phase fault/slip data. J. Struct. Geol. 28, 943–951.

Shan, Y., Fry, N., Lisle, R.J., 2009. Graphical construction for the direction of shear. J. Struct. Geol. 31, 476–478.

Shanley, R.J., Mahtab, M.A., 1976. Delineation and analysis of clusters in orientation data. Math. Geol. 8, 9–23.

Sibson, R.H., 1990. Conditions for fault-valve behaviour. In: Knipe, R.J., Rutter, E.H. (Eds.), Deformation Mechanisms, Rheology and Tectonics. vol. 54. Geological Society, London, Special Publications, pp. 15–28.

Simón, J.L., 2019. Forty years of paleostress analysis: has it attained maturity? J. Struct. Geol. 125, 124–133.

Simón-Gómez, J.L., 1986. Analysis of a gradual change in stress regime (example from eastern Iberian Chain, Spain). Tectonophysics 124, 37–53.

Sperner, B., Zweigel, P., 2010. A plea for more caution in fault-slip analysis. Tectonophysics 482, 29–41.

Sperner, B., Ratschbacher, L., Ott, R., 1993. Fault-striae analysis: a turbo pascal program package for graphical presentation and reduced stress tensor calculation. Comput. Geosci. 19, 1361–1388.

Steketee, J., 1958a. On volterra's dislocations in a semi-infinite elastic medium. Can. J. Phys. 36, 192–205.

Steketee, J., 1958b. Some geophysical applications of the elasticity theory of dislocations. Can. J. Phys. 36, 1168–1198.

Stephens, T.L., Walker, R.J., Healy, D., Bubeck, A., England, R.W., 2018. Mechanical models to estimate the paleostress state from igneous intrusions. Solid Earth 9, 847–858.

Storetvedt, K.M., 1987. Major late Caledonian and Hercynian shear movements on the Great Glen Fault. Tectonophysics 143, 253–267.

Tapponnier, P., Molnar, P., 1975. Cenozoic tectonics of Asia: effects of a continental collision. Science 189, 419–426.

Tchalenko, J.S., 1970. Similarities between shear zones of different magnitudes. Geol. Soc. Am. Bull. 81, 1625–1640.

Tiercelin, J.J., Chorowicz, J., Bellon, H., Richert, J.P., Mwambene, J.T., Walgenwitz, F., 1988. East Africa rift system: offset, age, and tectonic significance of the Tanganyika–Rukwa–Malawi intracontinental transcurrent fault zone. Tectonophysics 148, 241–252.

Timoshenko, S., 1953. History of Strength of Materials. McGraw-Hill, New York.

Tourneret, C., Laurent, P., 1990. Paleo-stress orientations from calcite twins in the North Pyrenean fore-land, determined by the Etchecopar inverse method. Tectonophysics 180, 287–302.

Toussaint, R., Aharonov, E., Koehn, D., Gratier, J.-P., Ebner, M., Baud, P., Rolland, A., Renard, F., 2018. Stylolites: a review. J. Struct. Geol. 114, 163–195.

Tranos, M.D., 2015. TR method (TRM): a separation and stress inversion method for heterogeneous fault-slip data driven by Andersonian extensional and compressional stress regimes. J. Struct. Geol. 79, 57–74.

Tranos, M.D., 2017. The use of stress tensor discriminator faults in separating heterogeneous fault-slip data with best-fit stress inversion methods. J. Struct. Geol. 102, 168–178.

Turner, F.J., 1953. Nature and dynamic interpretation of deformation lamellae in calcite of three marbles. Am. J. Sci. 251, 276–298.

Turner, F.J., Weiss, L.E., 1963. Structural Analysis of Metamorphic Tectonites. McGraw-Hill, New York.

Tuttle, O.F., 1949. Structural petrology of planes of liquid inclusions. J. Geol. 57, 331–356.

Twiss, R.J., Moores, E.M., 2001. Structural Geology. 7th printing of the 1992 ed. W. H. Freeman, New York.

Twiss, R.J., Moores, E.M., 2007. Structural Geology, second ed. W. H. Freeman, New York.

Twiss, R.J., Unruh, J.R., 1998. Analysis of fault-slip inversions: do they constrain stress or strain rate? J. Geophys. Res. 103 (B6), 12205–12222.

Twiss, R.J., Protzman, G.M., Hurst, S.D., 1991. Theory of slickenline patterns based on the velocity gradient tensor and microrotation. Tectonophysics 186, 215–239.

Twiss, R.J., Souter, B.J., Unruh, J.R., 1993. The effect of block rotations on the global seismic moment tensor and the patterns of seismic P and T axes. J. Geophys. Res. 98 (B1), 645–674.

van den Bos, A., 2007. Parameter Estimation for Scientists and Engineers. Wiley, Hoboken.

van der Pluijm, B.A., Marshak, S., 2004. Earth Structure: An Introduction to Structural Geology and Tectonics, second ed. WW Norton, New York.

van der Pluijm, B.A., Craddock, J.P., Graham, B.R., Harris, J.H., 1997. Paleostress in cratonic North America: implications for deformation of continental interiors. Science 277, 796.

Volterra, V., 1907. Sur l'équilibre des corps élastiques multiplement connexes. Ann. Sci. École Norm. Sup. 24, 401–517.

von Terzaghi, K., 1936. The shearing resistance of saturated soils and the angle between the planes of shear. In: First Int. Conf. Soil Mech. vol. 1. Harvard University, pp. 54–56.

Wallace, R.E., 1951. Geometry of shearing stress and relation to faulting. J. Geol. 59, 118–130.

Will, T.M., Powell, R., 1991. A robust approach to the calculation of paleostress fields from fault plane data. J. Struct. Geol. 13, 813–821.

Wise, D.U., 1964. Microjointing in basement, middle Rocky Mountains of Montana and Wyoming. GSA Bull. 75, 287–306.

Worum, G., van Wees, J.-D., Bada, G., van Balen, R.T., Cloetingh, S., Pagnier, H., 2004. Slip tendency analysis as a tool to constrain fault reactivation: a numerical approach applied to three-dimensional fault models in the Roer Valley rift system (southeast Netherlands). J. Geophys. Res. Solid Earth 109.

Xu, P., 2004. Determination of regional stress tensors from fault–slip data. Geophys. J. Int. 157, 1316–1330.

Yamaji, A., 2000. The multiple inverse method: a new technique to separate stresses from heterogeneous fault-slip data. J. Struct. Geol. 22, 441–452.

Yamaji, A., 2003. Are the solutions of stress inversion correct? Visualization of their reliability and the separation of stresses from heterogeneous fault-slip data. J. Struct. Geol. 25, 241–252.

Yamaji, A., 2007. Determination of stress from faults. In: An Introduction to Tectonophysics: Theoretical Aspects of Structural Geology. Terrapub, Tokyo (Chapter 11).

Yamaji, A., 2015. How tightly does calcite e-twin constrain stress? J. Struct. Geol. 72, 83–95.

Yamaji, A., 2016. Genetic algorithm for fitting a mixed Bingham distribution to 3D orientations: a tool for the statistical and paleostress analyses of fracture orientations. Island Arc 25, 72–83.

Yamaji, A., Sato, K., 2011. Clustering of fracture orientations using a mixed Bingham distribution and its application to paleostress analysis from dike or vein orientations. J. Struct. Geol. 33, 1148–1157.

Yamaji, A., Otsubo, M., Sato, K., 2006. Paleostress analysis using the Hough transform for separating stresses from heterogeneous fault-slip data. J. Struct. Geol. 28, 980–990. New Dynamics in Palaeostress Analysis.

Yamaji, A., Sato, K., Tonai, S., 2010. Stochastic modeling for the stress inversion of vein orientations: paleostress analysis of Pliocene epithermal veins in southwestern Kyushu, Japan. J. Struct. Geol. 32, 1137–1146.

Yin, Z.M., Ranalli, G., 1993. Determination of tectonic stress field from fault slip data: toward a probabilistic model. J. Geophys. Res. 98 (B7), 12165–12176.

Yin, Z.-M., Ranalli, G., 1995. Estimation of the frictional strength of faults from inversion of fault-slip data: a new method. J. Struct. Geol. 17, 1327–1335.

Žalohar, J., Vrabec, M., 2007. Paleostress analysis of heterogeneous fault-slip data: the Gauss method. J. Struct. Geol. 29, 1798–1810.

Zoback, M.D., 2007. Reservoir Geomechanics. Cambridge University Press.

Index

Note: Page numbers followed by *f* indicate figures, *t* indicate tables, and *np* indicate footnotes.

Printed in the United States
by Baker & Taylor Publisher Services